GCSE 9-1

HIGHER

MATHEMATICS

AQA

EXAM PRACTICE

Stephen Doyle

Author Stephen Doyle
Editorial team Haremi Ltd
Series designers emc design ltd
Typesetting York Publishing Solutions Pvt. Ltd.
Illustrations York Publishing Solutions Pvt. Ltd.
App development Hannah Barnett, Phil Crothers and Haremi Ltd

Designed using Adobe InDesign
Published by Scholastic Education, an imprint of Scholastic Ltd, Book End, Range Road, Witney, Oxfordshire, OX29 0YD
Registered office: Westfield Road, Southam, Warwickshire CV47 0RA
www.scholastic.co.uk

Printed by Bell & Bain Ltd, Glasgow
© 2017 Scholastic Ltd
1 2 3 4 5 6 7 8 9   7 8 9 0 1 2 3 4 5 6

British Library Cataloguing-in-Publication Data
A catalogue record for this book is available from the British Library.
ISBN 978-1407-16904-0

Notes from the publisher
Please use this product in conjunction with the official specification and sample assessment materials. Ask your teacher if you are unsure where to find them.

The marks and star ratings have been suggested by our subject experts, but they are to be used as a guide only.

Answer space has been provided, but you may need to use additional paper for your workings.

Contents

Topic 4

GEOMETRY AND MEASURES

Topic 5

PROBABILITY

Topic 6

STATISTICS

How to use this book

This Exam Practice Book has been produced to help you revise for your 9–1 GCSE in AQA Higher Mathematics. Written by an expert and packed full of exam-style questions for each subtopic, along with full practice papers, it will get you exam ready!

The best way to retain information is to take an active approach to revision. Don't just read the information you need to remember – do something with it! Transforming information from one form into another and applying your knowledge will ensure that it really sinks in. Throughout this book you'll find lots of features that will make your revision practice an active, successful process.

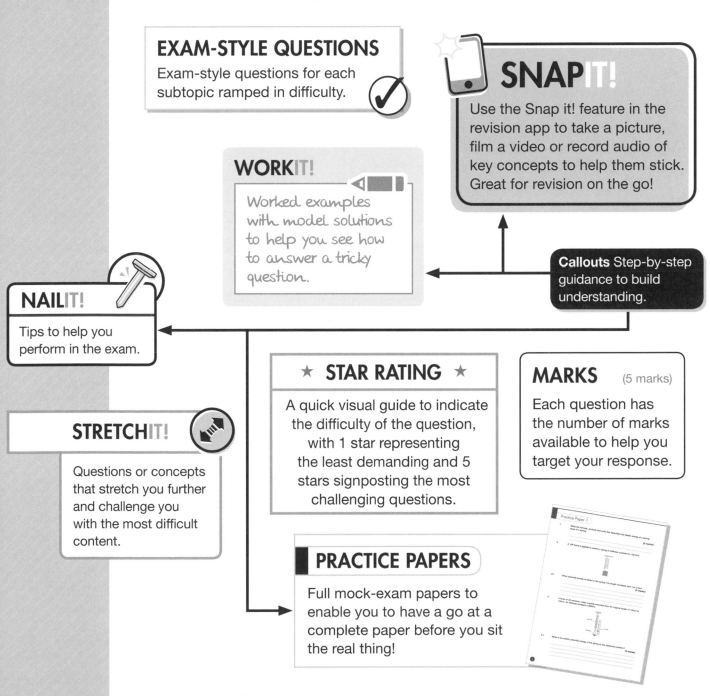

EXAM-STYLE QUESTIONS

Exam-style questions for each subtopic ramped in difficulty.

SNAP IT!

Use the Snap it! feature in the revision app to take a picture, film a video or record audio of key concepts to help them stick. Great for revision on the go!

WORK IT!

Worked examples with model solutions to help you see how to answer a tricky question.

Callouts Step-by-step guidance to build understanding.

NAIL IT!

Tips to help you perform in the exam.

STRETCH IT!

Questions or concepts that stretch you further and challenge you with the most difficult content.

★ STAR RATING ★

A quick visual guide to indicate the difficulty of the question, with 1 star representing the least demanding and 5 stars signposting the most challenging questions.

MARKS (5 marks)

Each question has the number of marks available to help you target your response.

PRACTICE PAPERS

Full mock-exam papers to enable you to have a go at a complete paper before you sit the real thing!

Use the AQA Higher Mathematics Revision Guide alongside the Exam Practice Book for a complete revision and practice solution. Written by a subject expert to match the new specification, the Revision Guide uses an active approach to revise all the content you need to know!

HOW TO REVISE!

PLAN YOUR REVISION

Get ahead by planning your revision!

Work out the **time** you have available for revising.

Think about when you work at your best. Are you a morning or an evening person?

Allocate **MORE TIME** for the topics you struggle with.

Revision works best in **SMALL BURSTS**, so keep sessions **SHORT AND SWEET**!

Remember to allow time to **PRACTISE** applying what you have revised.

Use your **revision app** to put together a revision timetable.

LOOK AFTER YOURSELF

Help your brain by looking after your whole body!

Take regular **breaks** from revising – your brain needs time to digest information in order to retain it.

HOTEL

Keep **hydrated** by drinking plenty of water – dehydration stops your brain from working at its full capacity.

Regular **exercise** helps stimulate the brain and will help you relax.

Get plenty of **sleep**, especially the night before an exam.

EAT WELL and limit unhealthy snacks – your brain needs fuel for memory and concentration.

Find methods of **relaxation** that work for you throughout the revision period.

BE PREPARED!

Limit potential stress on the day of an exam by getting everything you need ready the night before.

30

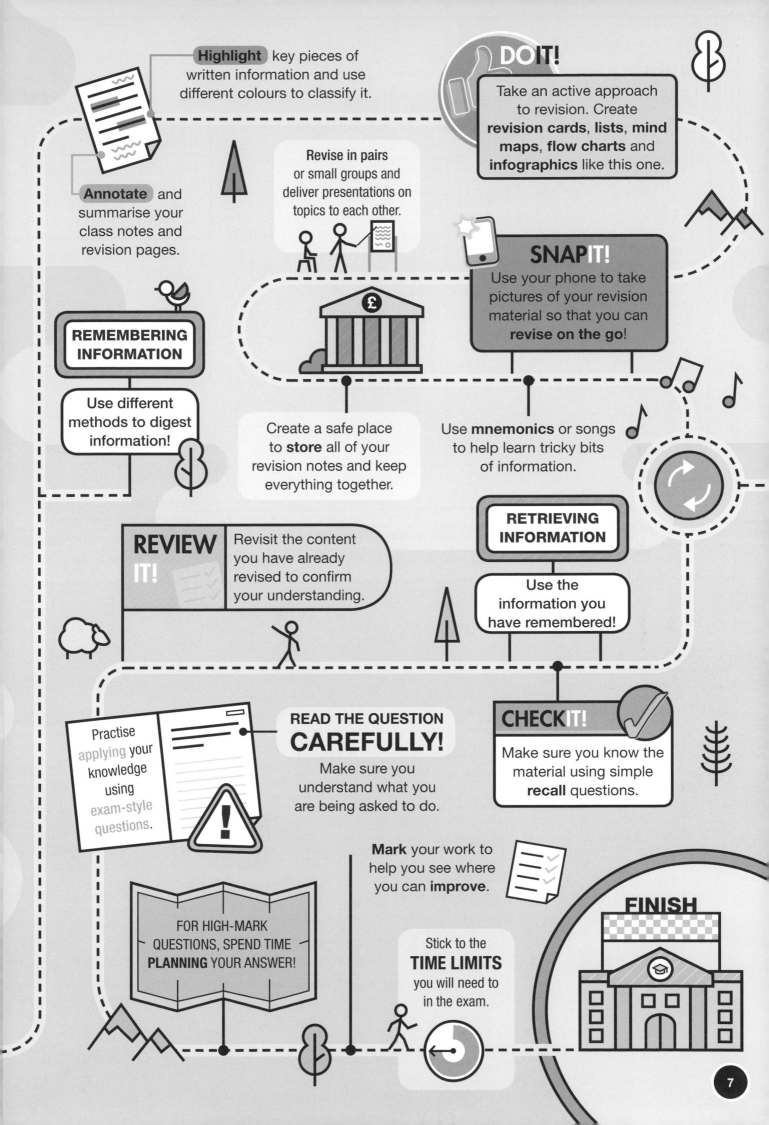

Highlight key pieces of written information and use different colours to classify it.

Annotate and summarise your class notes and revision pages.

Revise in pairs or small groups and deliver presentations on topics to each other.

DO IT!

Take an active approach to revision. Create **revision cards**, **lists**, **mind maps**, **flow charts** and **infographics** like this one.

SNAP IT!

Use your phone to take pictures of your revision material so that you can **revise on the go!**

REMEMBERING INFORMATION

Use different methods to digest information!

Create a safe place to **store** all of your revision notes and keep everything together.

Use **mnemonics** or songs to help learn tricky bits of information.

RETRIEVING INFORMATION

Use the information you have remembered!

REVIEW IT!

Revisit the content you have already revised to confirm your understanding.

Practise *applying* your knowledge using *exam-style questions*.

READ THE QUESTION CAREFULLY!

Make sure you understand what you are being asked to do.

CHECK IT!

Make sure you know the material using simple **recall** questions.

Mark your work to help you see where you can **improve**.

FOR HIGH-MARK QUESTIONS, SPEND TIME **PLANNING** YOUR ANSWER!

Stick to the **TIME LIMITS** you will need to in the exam.

FINISH

Number
Integers, decimals and symbols

NAILIT!

Make sure you understand terms such as integer, place value, ascending and descending.

(1) Arrange these numbers in descending order. (2 marks, ★)

$(-1)^3$ $(0.1)^2$ $\dfrac{1}{1000}$ 0.1 $\dfrac{1}{0.01}$

....

(2) Without using a calculator, work out (★)

a 0.035×1000 (1 mark) ....

b $12.85 \div 1000$ (1 mark) ....

c $(-3) \times 0.09 \times 1000$ (1 mark) ....

d $(-1) \times (-0.4) \times 100$ (1 mark) ....

[Total: 4 marks]

(3) $0.86 \times 54 = 46.44$

Without using a calculator, work out (★★) ◄———— Think about place value.

a 86×54 (1 mark)   **c** $\dfrac{4644}{54}$ (1 mark)

....

b 8.6×540 (1 mark)   **d** $\dfrac{46.44}{0.086}$ (1 mark)

....

[Total: 4 marks]

(4) Place the correct symbol from the following list in the box.
The symbols can be used once, more than once or not at all. (★★)

$<$ $\geq$ $\leq$ $=$ $>$

a 12.56×3.45 ☐ 0.1256×345 (1 mark) ◄———— Work out the calculation either side of the box and then insert the correct symbol in the box.

b $(-8)^2$ ☐ -64 (1 mark)

c $6 - 12$ ☐ $8 - 14$ (1 mark)

d $(-7) \times (0)$ ☐ $(-7) \times (-3)$ (1 mark)

[Total: 4 marks]

Addition, subtraction, multiplication and division

Do **not** use a calculator for any of these calculations.

① Work out (★★)

a 67.78 + 8.985 (1 mark)   c 93.1 − 1.77 (1 mark)

....

b 124.706 + 76.9 + 0.04   d 23.7 + 8.94 − 22.076
(1 mark)   (1 mark)

....

② Work out (★★)

a 147 × 8 (1 mark)   c 9.7 × 4.6 (1 mark)   e 486 ÷ 18 (1 mark)

....

b 57 × 38 (1 mark)   d 1.24 × 0.53 (1 mark)   f 94.5 ÷ 1.5 (1 mark)

....

[Total: 6 marks]

③ Work out (★★★)

a 34^2 (1 mark)   b $\dfrac{1.5 \times 2.5}{0.5}$ (1 mark)   c 2.4^2 (1 mark)

....

[Total: 3 marks]

NAILIT!

On Paper 1 these calculations must be done without a calculator.

WORKIT!

Work out 23.48 − 8.362.

```
  ¹2¹3 . 4 ⁷8 ¹0
-    8 . 3 6 2
  ─────────────
  1 5 . 1 1 8
```

[Total: 4 marks]

WORKIT!

Work out 8.97 ÷ 1.3.

Multiply the numerator and denominator by a factor of 10 to make the denominator a whole number.

$$\frac{8.97}{1.3} = \frac{89.7}{13}$$

```
        6 . 9
   13) 8 9 . 7
       7 8
       ─────
       1 1 7
       1 1 7
       ─────
           0
```

9

Using fractions

NAILIT!

Make sure you understand the terms numerator, denominator, improper fraction and mixed number.

(1) Fill in the missing numbers in these equivalent fractions. (1 mark, ★)

$$\frac{2}{5} = \frac{16}{\square} = \frac{\square}{75} = \frac{50}{\square}$$

(2) Simplify these fractions. Give your answers as mixed numbers in their simplest form. (★★)

a $\frac{64}{12}$ (1 mark) ....

When calculating with fractions, always remember to cancel as far as possible.

b $\frac{124}{13}$ (1 mark) ....

[Total: 2 marks]

(3) Work out these calculations, simplifying your answer if possible. (★★)

a $4\frac{1}{4} \times 1\frac{2}{3}$ (1 mark)

SNAPIT! Using fractions

If you multiply or divide the top and bottom of a fraction by the same number, the size of the fraction stays the same.

....

b $1\frac{7}{8} \div \frac{1}{4}$ (1 mark)

....

c $3\frac{1}{5} - \frac{3}{4}$ (1 mark)

Convert mixed numbers to improper fractions first.

....

[Total: 3 marks]

(4) In a class of students, all the students walk, cycle, travel by bus or travel by car to school. $\frac{2}{7}$ of the students walk, $\frac{3}{8}$ of the students cycle and $\frac{1}{4}$ travel by bus. What fraction of the students travel to school by car? (2 marks, ★)

....

(5) Put the following list of fractions into order of size starting with the smallest. (2 marks, ★★)

$$\frac{2}{3} \qquad \frac{3}{4} \qquad \frac{7}{8} \qquad \frac{1}{2} \qquad \frac{7}{12}$$

NAILIT!

Change the fractions so they have the same denominator (i.e. bottom number in the fraction).

....

Different types of number

1 Use the numbers from the list to answer the following questions. (★)

2   7   49   6   14

Write down the number that is

a a factor of 77 (1 mark)

....

b a square number (1 mark)

....

c an even prime number (1 mark)

....

d not a factor of 98 (1 mark)

....

e a multiple of 3. (1 mark)

....

NAILIT!

Make sure you understand terms such as multiple, factor, common factor, highest common factor, lowest common multiple and prime.

STRETCHIT!

A **perfect number** is one whose factors add up to the number itself. An **abundant number** has factors that add up to more than the number itself. Write a factor tree for the number 216. Is it one of these?

[Total: 5 marks]

WORKIT!

The number 540 can be written as the product of its prime factors:

$540 = 2^2 \times 3^3 \times 5$

a Write 168 as the product of its prime factors.

$168 = 2 \times 84$

$= 2 \times 2 \times 42$ ← All the numbers in the product must be prime. If they aren't, keep breaking the number down until they are.

$= 2 \times 2 \times 6 \times 7$

$= 2 \times 2 \times 2 \times 3 \times 7$

$= 2^3 \times 3 \times 7$

b Work out the highest common factor of 540 and 168.

Both numbers have 2^2 and 3 in their prime factors, ← $2^3 = 2^2 \times 2$
so highest common factor = $2^2 \times 3 = 12$

c Work out the lowest common multiple of 540 and 168.

Multiplying the prime factors of both, ignoring duplicates, the lowest common multiple = $2^3 \times 3^3 \times 5 \times 7 = 7560$

(2) The number 945 can be written as the product of prime factors: $945 = 3^3 \times 5 \times 7$. (★★)

a Write the number 693 as the product of prime factors. (2 marks)

....

b Work out the highest common factor of 945 and 693. (1 mark)

....

c Work out the lowest common multiple of 945 and 693. (1 mark)

....

[Total: 4 marks]

(3) Find the lowest common multiple of 49 and 63. (2 marks, ★★)

....

(4) Tom rings a bell every 15 seconds, Julie bangs a drum every 20 seconds and Phil hits a triangle every 25 seconds. If they all sound a note together at the same time, after how many minutes will they next make a note at the same time? (2 marks, ★★★)

> Start by finding the product of prime factors for each time. Then find the lowest common multiple.

.... minutes

Listing strategies

NAILIT!

Sometimes it's hard to spot a method for solving a question. Making a list can make it easier to see what to do next.

(1) A frog croaks every 14 seconds and a parrot squawks every 15 seconds.

Assuming that they both make their sound together initially, what is the minimum amount of time in seconds before they will sound together again? (2 marks, ★★)

Write down all the multiples of the times to find the first time which is common to both lists.

.... seconds

(2) Dhaya has some chocolates which she shares with her friends.

If she gives 6 chocolates to each friend she will have one chocolate left.

If she gave each friend 7 sweets she would need another 4 sweets.

Start by listing the multiples of 6 and 7.

By producing lists, find the number of friends she could have. (3 marks, ★★)

....

(3) The ratio of male to female students in a college is 5 : 6.
There are 100 more female students than male students in the college.
Find the total number of students in the college. (2 marks, ★★)

....

(4) A group of 6 students in a primary school are asked to work in pairs on a science project.
How many possible pairs could there be? (2 marks, ★★★)

Represent each pupil by a letter and write down the possible pairs excluding any pairs that are the same.

....

The order of operations in calculations

Do **not** use a calculator for any of these calculations.

(1) Ravi is finding the value of this expression. (★)

$24 \div 3 + 8 \times 4$.

> Use BIDMAS: brackets, indices (powers and roots), division, multiplication, addition, subtraction.

Here is his working out.

$$24 \div 3 + 8 \times 4 = 8 + 8 \times 4$$
$$= 16 \times 4$$
$$= 64$$

Ravi's answer is wrong.

> **NAILIT!**
>
> Do not work out calculations in the order they appear from left to right. Use BIDMAS.

a Explain what he has done wrong. (3 marks)

....

....

b Showing your working, what is the correct answer? (1 mark)

....

[Total: 4 marks]

(2) Evaluate (★★)

a $9 \times 7 \times 2 - 24 \div 6$ (1 mark)

....

b $2 - (-27) \div (-3) + 4$ (1 mark)

....

c $(4 - 16)^2 \div 4 + 32 \div 8$ (1 mark)

....

[Total: 3 marks]

(3) Evaluate (★★)

a $1 + 4 \div \frac{1}{2} - 3$ (1 mark)

....

b $15 - (1 - 2)^2$ (1 mark)

....

c $\sqrt{4 \times 12 - 2 \times (-8)}$ (1 mark)

....

[Total: 3 marks]

Indices

Do **not** use a calculator for any of these calculations.

1. Express each of these as a single power of 10. (★★)

   a $10^5 \times 10$ (1 mark)

   c $\dfrac{10^5 \times 10^3}{10^2}$ (1 mark)

   b $(10^4)^2$ (1 mark)

   d $(10^6)^{\frac{1}{2}}$ (1 mark)

> **NAILIT!**
>
> Make sure that you learn the rules of indices.

> **NAILIT!**
>
> If a number appears on its own without a power then the power to which it is raised is 1. So 10 can be written as 10^1.

[Total: 4 marks]

2. Without using a calculator, write the value of (★★★)

   a 5^0 (1 mark)

   c $8^{\frac{1}{3}}$ (1 mark)

   b 3^{-2} (1 mark)

   d $49^{\frac{1}{2}}$ (1 mark)

[Total: 4 marks]

3. Without using a calculator, work out (★★★★)

   a $\left(\dfrac{8}{27}\right)^{-\frac{1}{3}}$ (2 marks)

   c $36^{-\frac{1}{2}}$ (2 marks)

   b $\left(\dfrac{1}{4}\right)^{-2}$ (2 marks)

   d $16^{\frac{3}{2}}$ (2 marks)

[Total: 8 marks]

4. Find the value of x such that $5^{2x} = 125$. (2 marks, ★★★)

Surds

① Evaluate (★★★)

    a $\left(\sqrt{5}\right)^2$ (1 mark)   **b** $3\sqrt{2} \times 5\sqrt{2}$ (1 mark)   **c** $\left(3\sqrt{2}\right)^2$ (2 marks)

NAILIT!

The square of a square root is the number inside the square root, e.g. $\left(\sqrt{2}\right)^2 = 2$.

....

[Total: 4 marks]

② Rationalise the denominator and simplify $\dfrac{15}{4\sqrt{3}}$ (2 marks, ★★★★)

NAILIT!

'Rationalise' means to remove the surd from the denominator. To do this you multiply both numerator and denominator by that surd.

....

③ Show that $\left(2 + \sqrt{3}\right)\left(2 - \sqrt{3}\right)$ simplifies to 1. (2 marks, ★★★★)

④ $\sqrt{3}\left(3\sqrt{24} + 4\sqrt{6}\right)$ can be simplified to $a\sqrt{2}$. Find the value of a. (3 marks, ★★★★) ◄

Multiply every term in the bracket by the term outside the bracket.

....

⑤ Work out $\left(1 - \sqrt{5}\right)\left(3 + 2\sqrt{5}\right)$. (2 marks, ★★★★) ◄

Multiply every term in the first bracket by every term in the second bracket.

....

⑥ Show that $\dfrac{1}{\sqrt{2}} + \dfrac{1}{4}$ can be written as $\dfrac{1 + 2\sqrt{2}}{4}$ (3 marks, ★★★★)

NAILIT!

To remove the surd in, for example, $\dfrac{5}{1-\sqrt{3}}$ multiply the numerator and denominator by $1 + \sqrt{3}$.

⑦ Show that $\dfrac{2}{1-\frac{1}{\sqrt{2}}}$ can be written as $4 + 2\sqrt{2}$. (3 marks, ★★★★★)

⑧ Show that $\dfrac{3}{\sqrt{3}} + \sqrt{75} + \left(\sqrt{2} \times \sqrt{6}\right) = 8\sqrt{3}$. (3 marks, ★★★★)

Standard form

① Write these numbers in standard form. (★)

a 0.00255 (1 mark)

....

c 0.000000089 (1 mark)

....

b 10 060 000 000 (1 mark)

....

[Total: 3 marks]

② Without using a calculator, work out these calculations. Give your final answer in standard form. (★★★★)

a $3 \times 10^6 \times 2 \times 10^8$ (1 mark) ◄ *Multiply the number parts. Multiply the powers of 10 using the index laws.*

....

b $5.5 \times 10^{-3} \times 2 \times 10^8$ (2 marks)

....

c $\dfrac{8 \times 10^5}{4 \times 10^3}$ (1 mark)

....

d $\dfrac{5 \times 10^{-3}}{0.5}$ (2 marks)

....

If necessary, change the decimal point in the number and the power of 10 to give the answer in standard form.

e $\dfrac{4.5 \times 10^{-6}}{0.5 \times 10^{-3}}$ (1 mark) ◄

....

WORKIT!

Without using a calculator, work out, giving your final answer in standard form

a $3.5 \times 10^{-4} \times 4 \times 10^8$

$3.5 \times 10^{-4} \times 4 \times 10^8$

$= (3.5 \times 4) \times (10^{-4} \times 10^8)$

$= 14 \times 10^4$

$= 1.4 \times 10^5$

b $(4.5 \times 10^{-2}) \div (6 \times 10^{-5})$

$(4.5 \times 10^{-2}) \div (6 \times 10^{-5})$

$= (4.5 \div 6) \times (10^{-2} \div 10^{-5})$

$= 0.75 \times 10^3$

$= 7.5 \times 10^2$

[Total: 7 marks]

③ Without using a calculator, work out

$2.4 \times 10^3 + 2.8 \times 10^2$

Give your answer as an ordinary number. (2 marks, ★★★)

....

④ If $a \times 10^4 + a \times 10^2 = 33330$, find the value of a. (3 marks, ★★★★)

....

Converting between fractions and decimals

① Without using a calculator, convert these fractions into decimals. (★)

a $\frac{11}{20}$ (1 mark)

b $\frac{3}{8}$ (1 mark)

NAILIT!

Check that you know all the common fractions as decimal conversions (e.g. $\frac{1}{5} = 0.2$).

.... **[Total: 2 marks]**

② Without using a calculator, state whether each of these fractions will produce a terminating or a recurring decimal. (★★)

a $\frac{1}{16}$ (1 mark)

....

b $\frac{5}{7}$ (1 mark)

....

c $\frac{13}{35}$ (1 mark)

STRETCHIT!

Look at the prime factors in the denominator. If they are all 2 and/or 5, then the decimal will terminate.

.... **[Total: 3 marks]**

③ Prove that the recurring decimal $0.\dot{4}0\dot{2}$ can be written as the fraction $\frac{134}{333}$ (3 marks, ★★★★)

Start by letting x be equal to the recurring decimal.

WORKIT!

Write $0.\dot{7}\dot{1}$ as a fraction.

$$\text{Let } x = 0.717171...$$
$$100x = 71.717171...$$
$$100x - x = 71$$
$$99x = 71$$
$$x = \frac{71}{99}$$

④ Write the recurring decimal $0.6\dot{5}\dot{2}$ as a fraction in its simplest form. (3 marks, ★★★★)

Remember to cancel the resulting fraction.

....

Converting between fractions and percentages

(1) Convert these percentages to fractions in their simplest form. (★)

a 35% (1 mark)

c 76% (1 mark)

....

....

b 7% (1 mark)

d 12.5% (1 mark)

Write the percentage as a fraction with denominator 100. Simplify as far as possible.

....

....

[Total: 4 marks]

(2) Without using a calculator, write each of these fractions as percentages. (★★)

a $\frac{1}{5}$ (1 mark)

c $\frac{150}{60}$ (1 mark)

....

....

Where the denominator is a factor of 100, multiply both the top and the bottom of the fraction by the amount needed to get 100 on the bottom.

b $\frac{17}{25}$ (1 mark)

d $\frac{7}{40}$ (1 mark)

....

....

[Total: 4 marks]

(3) Convert the fraction $\frac{8}{15}$ to a percentage. Give your answer to 2 decimal places. (2 marks, ★★)

....

(4) Jake scored $\frac{66}{90}$ in a physics test and 75% in a chemistry test. Show, with reasons, which test he did better in. (2 marks, ★★)

NAILIT!

Where the denominator is not a factor of 100, divide the numerator by the denominator and multiply by 100.

Fractions and percentages as operators

(1) Calculate 70% of £49.70. (1 mark, ★)

....

(2) In a batch of apples, 8% of the apples are bad. If there are 600 apples in the batch, how many of them are bad? (2 marks, ★)

The word 'of' in these questions means multiply.

....

(3) In a batch of car parts, 12% of them were rejected. How many car parts are likely to be accepted in a batch of 8000? (2 marks, ★★)

....

(4) Chloe buys a new car costing £12 000 + VAT at 20%. (★★)

a Calculate the total cost of the car including VAT. (1 mark)

....

b Chloe has to pay the garage 20% of the total cost of the car as a deposit and the remainder is paid over 36 equal monthly payments. Calculate her monthly payment. (2 marks)

....

[Total: 3 marks]

(5) In a class of students, $\frac{2}{3}$ are male. Out of the males $\frac{4}{11}$ are taking chemistry.

What fraction of the class are males **not** taking chemistry? (2 marks, ★★★)

....

Standard measurement units

① Convert 1.75 km to centimetres. (1 mark, ★)

.... cm

NAILIT!

Learn the conversion factors for units of length, mass, capacity and time.

② A carton contains 3 litres of juice. How many complete 175 ml glasses could be filled from the carton? (2 marks, ★★)

....

NAILIT!

Always ask yourself before you start 'is the answer going to be bigger or smaller than the number in the question?'.

③ A children's paddling pool is to be filled with 900 litres of water. The only container they can find to fill the paddling pool has a volume of 700 cm³.

First change the volumes into the same units.

How many containers full of water would be required to fill the paddling pool? Give your answer to the nearest whole number. (3 marks, ★★)

....

④ 12 g of carbon contains 6.02×10^{23} atoms. (★★★★)

a Find the mass of 1 atom of carbon. Give your answer in standard form in g, correct to 3 significant figures. (2 marks)

.... g

b Give your answer to part a in kg, also in standard form and correct to 3 significant figures. (2 marks)

.... kg

[Total: 4 marks]

⑤ An electron has a mass of 9.11×10^{-31} kg. In an atom of gold there are 79 electrons. Work out the mass of electrons in one atom of gold. Give your answer in standard form to 3 significant figures in grams. (3 marks, ★★★★)

.... g

Rounding numbers

1 Write these numbers correct to the nearest whole number. (★)

  a 34.7 (1 mark)

....

  b 100.6 (1 mark)

....

  c 0.078 (1 mark)

....

  d 0.4999 (1 mark)

....

  e 1.5001 (1 mark)

....

> ## WORKIT!
>
> Round 6.785 to 2 decimal places.
>
> **1** Look at the 3rd digit after the decimal place: 5.
>
> **2** If is 5 or more, round up; if it is less than 5, round down.
>
> 6.785 = 6.79 (to 2 d.p.)

[Total: 5 marks]

2 The number 34.8765 is given to 4 decimal places. (★)

Write this number to

  a 2 decimal places (1 mark)   **b** 3 decimal places. (1 mark)

[Total: 2 marks]

....

3 Write each number to the specified number of significant figures. (★)

  a 12758 (3 s.f.) (1 mark)

....

  b 0.01055 (2 s.f.) (1 mark)

....

  c 7.46×10^{-5} (1 s.f.) (1 mark)

....

> ## WORKIT!
>
> Round 0.078366 to 3 significant figures.
>
> **1** Identify the first significant figure: 7.
>
> **2** Count the required number of figures: 0.0783.
>
> **3** Round the next digit: 0.0784.
>
> 0.078366 = 0.0784 (to 3 s.f.)

[Total: 3 marks]

4 Work out these calculations. Give all your answers to 3 significant figures. (★★)

  a $(0.18 \times 0.046)^2 - 0.01$ (1 mark)

....

  b $\dfrac{1200 \times 1.865}{2.6 \times 25}$ (1 mark)

....

  c $\dfrac{36}{0.07} \times 12 \div \dfrac{1}{2}$ (1 mark)

....

[Total: 3 marks]

Estimation

(1) Estimate the value of 4.6 × 9.8 × 3.1. Give your answer to 1 significant figure. (2 marks, ★)

NAILIT!

To estimate answers, write all the values in the calculation to 1 significant figure and give the answer to 1 significant figure.

....

(2) **a** Work out $19.87^2 - \sqrt{404} \times 7.89$.

Write down all the numbers on your calculator display. (2 marks, ★)

....

b Use estimation to check your answer to part a, showing all your working. (1 mark, ★)

....

[Total: 3 marks]

(3) Work out an estimate for $(0.52 \times 0.83)^2$. (2 marks, ★★)

....

(4) Work out an estimate for $\sqrt{5.08} + 4.10 \times 5.45$ (2 marks, ★★★)

....

WORKIT!

Estimate the value of $\sqrt{31}$.

1 Find the square root of the perfect squares that are above and below the value.

$\sqrt{25} = 5$ and $\sqrt{36} = 6$.

2 Take the value between them and square it to see whether this makes a good estimate.

5.5^2 is 30.25, just slightly less than 31. So 5.5 is a reasonable estimate.

(5) Estimate the value of $\sqrt{112}$. Give your answer to 1 decimal place. (2 marks, ★★★) ◄——————

Write the numbers above and below 112 that are perfect squares.

....

23

(6) Estimate the value of

$$\frac{0.89 \times 7.51 \times 19.76}{2.08 \times 5.44 \times 3.78}$$

Give your answer to 1 significant figure. (2 marks, ★★)

....

(7) An electron has a mass of 9.11×10^{-31} kg. An atom of carbon contains 6 electrons. (★★★★)

a Work out an estimate for the mass of the electrons in 100 carbon atoms. Give your answer correct to 1 significant figure in standard form in kg. (2 marks)

....kg

b Is your answer to part a an overestimate or underestimate? Give a reason for your answer. (1 mark)

....

....

....

....

[Total: 3 marks]

Upper and lower bounds

1 The mass of a brick, m, is 2.34 kg correct to 3 significant figures.

Write down the error interval for the mass of the brick. (1 mark, ★★)

....

2 The power, P, in watts produced in an electric resistor is given by the formula

$P = IV$

where I is the current in amps and V is the voltage in volts.

$P = 3.052$ correct to 4 significant figures

$I = 1.24$ amps correct to 3 significant figures (★★★★★)

a i Calculate the upper bound for V. (1 mark)

ii Calculate the lower bound for V. (1 mark)

b Using your answers for part a, work out the value of V to a suitable degree of accuracy. Give a reason for your answer. (2 marks)

....

[Total: 4 marks]

3 A bookcase has a shelf length of 1.1 m correct to 2 significant figures.

A maths textbook is 3.0 cm thick correct to 2 significant figures.

How many books will definitely fit onto the shelf? (3 marks, ★★★★)

....

WORKIT!

A length of a rod, l, is 48 cm to 2 significant figures.

Write down the error interval for the rod.

Upper bound = 48.5 cm

Lower bound = 47.5 cm

$47.5 \leq l < 48.5$ cm

NAILIT!

The error interval is the possible range of values and is written lower bound $\leq x <$ upper bound.

SNAPIT!

Upper and lower bounds

The upper bound is the maximum value a measurement can have. The lower bound is the minimum value.

NAILIT!

Rearrange the equation to make V the subject. Then identify the bounds that would give the fraction the greatest value and the lowest value.

Algebra
Simple algebraic techniques

(1) Identify whether each of these is a formula, expression, equation or identity. (★)

   a $v^2 = u^2 + 2as$ (1 mark)

   ....

   b $5x(2x + y) = 10x^2 + 5xy$ (1 mark)

   ....

   c $6a^2b$ (1 mark)

   ....

   d $(2a^2b)^2 = 4a^4b^2$ (1 mark)

   ....

   e $P = I^2R$ (1 mark)

   ....

[Total: 5 marks]

(2) Simplify $4x + 3x \times 2x - 3x$. (2 marks, ★★)

....

(3) Karl is trying to work out two values of y for which $y^3 - y = 0$.

The two values he finds are 1 and -1.

Are these two values correct? You must show your working. (2 marks, ★★★)

....

(4) Simplify these expressions. (★★)

   a $6x - (-4x)$
   (1 mark)

   b $x^2 - 2x - 4x + 3x^2$
   (1 mark)

   c $(-2x)^2 + 6x \times 3x - 4x^2$
   (2 marks)

   ....

[Total: 4 marks]

(5) $s = \dfrac{v^2 - u^2}{2a}$ (★★)

Work out the value of s when

   a $v = 3$, $u = 1$ and $a = 2$
   (1 mark)

   b $v = -4$, $u = 3$ and $a = 4$
   (1 mark)

   c $v = 5$, $u = -2$ and $a = -7$
   (1 mark)

   ....

[Total: 3 marks]

Removing brackets

① Expand the brackets for these expressions. (★)

  a $8(3x - 7)$ (1 mark)   **b** $-3(2x - 4)$ (1 mark)

.... **[Total: 2 marks]**

> **NAILIT!**
>
> Watch out for negative numbers outside the bracket as the signs will change when you multiply them out.

② Simplify (★★)

  a $3(2x - 1) - 3(x - 4)$ (2 marks) **c** $5ab(2a - b)$ (2 marks)

....

  b $4y(2x + 1) + 6(x - y)$ (2 marks) **d** $x^2y^3(2x + 3y)$ (2 marks)

> **NAILIT!**
>
> Use the laws of indices when you expand brackets.

.... **[Total: 8 marks]**

③ Expand and simplify (★★)

  a $(m - 3)(m + 8)$ (2 marks)   **c** $(3x - 1)^2$ (2 marks)

....

  b $(4x - 1)(2x + 7)$ (2 marks)   **d** $(2x + y)(3x - y)$ (3 marks)

.... **[Total: 9 marks]**

④ Expand and simplify (★★)

  a $(x + 5)(x + 2)$ (2 marks)   **c** $(x - 7)(x + 1)$ (2 marks)

....

  b $(x + 4)(x - 4)$ (2 marks)   **d** $(3x + 1)(5x + 3)$ (2 marks)

.... **[Total: 8 marks]**

⑤ Expand and simplify (★★★★)

  a $(x + 3)(x - 1)(x + 4)$ (3 marks)   **b** $(3x - 4)(2x - 5)(3x + 1)$ (3 marks)

.... **[Total: 6 marks]**

Factorising

(1) Factorise fully (★★)

   a $25x^2 - 5xy$ (2 marks)   **b** $4\pi r^2 + 6\pi x$ (2 marks)   **c** $6a^3b^2 + 12ab^2$ (2 marks)

....

[Total: 6 marks]

WORKIT!

Factorise $15xy + 3x^2$.

> Take the common factors outside the brackets.

$15xy + 3x^2 = 3x(5y + x)$

SNAPIT! Factorising

Factorising is the reverse process to expanding the brackets.

(2) Factorise (★★★)

   a $9x^2 - 1$ (2 marks)

....

   b $16x^2 - 4$ (2 marks)

....

> Use the difference of two squares:
> $a^2 - b^2 = (a + b)(a - b)$

[Total: 4 marks]

NAILIT!

Make sure that you take out all factors.

When a question says 'factorise fully', there is usually more than one factor. But if the question just says 'factorise' still check for more than one factor.

(3) Factorise (★★★)

   a $a^2 + 12a + 32$ (2 marks)   **b** $p^2 - 10p + 24$ (2 marks)

....

[Total: 4 marks]

WORKIT!

Factorise $x^2 - 2x - 8$.

> Find two numbers that multiply to make -8 and sum to make -2.

2 and -4

$x^2 - 2x - 8 = (x + 2)(x - 4)$

(4) Factorise (★★★)

   a $a^2 + 12a$ (2 marks)   **c** $x^2 - 11x + 30$ (2 marks)

....

   b $b^2 - 9$ (2 marks)

....   **[Total: 6 marks]**

(5) Factorise (★★★)

a $3x^2 + 20x + 32$ (2 marks)

c $2x^2 - x - 10$ (2 marks)

....

....

b $3x^2 + 10x - 13$ (2 marks)

....

[Total: 6 marks]

(6) Work out $\dfrac{x + 15}{2x^2 - 3x - 9} + \dfrac{3}{2x + 3}$.

Give your answer in its simplest form. (4 marks, ★★★★★)

....

(7) Write $\dfrac{1}{8x^2 - 2x - 1} \div \dfrac{1}{4x^2 - 4x + 1}$ in the form $\dfrac{ax + b}{cx + d}$ where a, b, c and d are integers.

(3 marks, ★★★★★)

....

Changing the subject of a formula

① Make T the subject of the formula $PV = nRT$. (2 marks, ★★)

....

② Make y the subject of the formula $2y + 4x - 1 = 0$. (2 marks, ★★)

....

③ Make a the subject of the formula $v = u + at$. (2 marks, ★★★)

....

④ Make x the subject of the formula $y = \frac{x}{5} - m$. (2 marks, ★★★)

....

⑤ Make v the subject of the formula $E = \frac{1}{2}mv^2$. (2 marks, ★★★)

....

⑥ The volume of a cone is given by the formula $V = \frac{1}{3}\pi r^2 h$ where V is the volume, r is the radius and h is the perpendicular height. (★★★)

a Rearrange the formula to make r the subject. (2 marks)

....

b Find the radius of a cone with a volume of 100 cm³ and a height of 8 cm.
Give your answer to 2 decimal places. (2 marks)

....

[Total: 4 marks]

(7) A straight line has the equation $y = 3x - 9$. (★★)

a Rearrange the equation to make x the subject. (1 mark)

....

b Find the value of x when $y = 3$. (1 mark)

....

[Total: 2 marks]

(8) Make x the subject of $3y - x = ax + 2$. (4 marks, ★★★★)

....

NAILIT!

When the subject appears on both sides, first get the terms containing the subject on one side. Then collect like terms or factorise to make sure the subject only appears once.

(9) **a** Make c the subject of the formula $c^2 = \dfrac{(16a^2\, b^4\, c^2)^{\frac{1}{2}}}{4a^2\, b}$.

(2 marks, ★★★★)

....

b $a = 2.8$ and $b = 3.2$, both to 1 decimal place. (3 marks, ★★★★★)

Work out the upper and lower bounds of c. Give your answers to 3 significant figures.

....

[Total: 5 marks]

Solving linear equations

1 Solve these equations. (★★)

a $2x + 11 = 25$ (1 mark)

....

b $3x - 5 = 10$ (1 mark)

....

c $15x = 60$ (1 mark)

....

d $\frac{x}{4} = 8$ (1 mark)

....

e $\frac{4x}{5} = 20$ (1 mark)

....

f $\frac{2x}{3} = -6$ (1 mark)

....

g $5 - x = 7$ (1 mark)

....

h $\frac{x}{7} - 9 = 3$ (1 mark)

....

> **NAILIT!**
>
> To solve a linear equation with only one unknown value perform the same operation(s) to both sides (add, subtract, multiply or divide) to get the unknown value on its own on one side.

[Total: 8 marks]

2 Solve the equation $5x - 1 = 2x + 1$. (2 marks, ★★)

....

3 Solve the equations. (★★★)

a $\frac{1}{4}(2x - 1) = 3(2x - 1)$ (3 marks)

....

b $5(3x + 1) = 2(5x - 3) + 3$ (3 marks)

....

[Total: 6 marks]

Solving quadratic equations using factorisation

WORKIT!

Solve $x^2 - 2x - 2 = 2x + 3$.

1 Rewrite the equation as a quadratic equal to zero.

$x^2 - 4x - 5 = 0$

2 Factorise.

$(x - 5)(x + 1) = 0$

3 Set each bracket equal to zero.

$x - 5 = 0$ or $x + 1 = 0$

$x = 5$ or $x = -1$

NAILIT!

Make sure you can expand brackets and factorise before attempting these questions.

NAILIT!

Remember that a quadratic must be equal to zero before you can factorise and hence solve it.

① **a** Factorise $x^2 - 7x + 12$. (2 marks, ★★★)

....

b Solve $x^2 - 7x + 12 = 0$. (1 mark, ★★★)

....

[Total: 3 marks]

② **a** Factorise $2x^2 + 5x - 3$.
(3 marks, ★★★★)

b Solve $2x^2 + 5x - 3 = 0$.
(1 mark, ★★★)

NAILIT!

You need to recognise quadratic expressions that can be factorised into two brackets.

....

[Total: 4 marks]

③ Solve the equation $x^2 - 3x - 20 = x - 8$. (4 marks, ★★★★)

....

④ a Show that the equation $x(x - 8) - 7 = x(5 - x)$ can be rearranged to $2x^2 - 13x - 7 = 0$. (2 marks, ★★★)

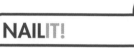

NAILIT!

If a quadratic equation appears in a problem, always check whether both solutions are possible or only one of them.

b Hence find the solutions to $x(x - 8) + 7 = x(5 - x)$. (3 marks, ★★★★)

....

[Total: 5 marks]

⑤ The diagram shows a trapezium with the sides measured in cm.

The trapezium has an area of $16\,cm^2$.

Find the value of x. (4 marks, ★★★★★)

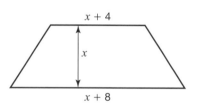

NAILIT!

Area of a trapezium $= \frac{1}{2}(a + b)h$ where a and b are the lengths of the two parallel sides and h is the distance between them.

....

Solving quadratic equations using the formula

(1) **a** Show that $\frac{3}{x+7} = \frac{2-x}{x+1}$ can be written as $x^2 + 8x - 11 = 0$.

(3 marks, ★★★★★)

....

b Hence solve the equation $\frac{3}{x+7} = \frac{2-x}{x+1}$.

Give your answers to 2 decimal places. (2 marks, ★★★★★)

NAILIT!

Quadratic equations of the form $ax^2 + bx + c = 0$ $(a \neq 0)$ can be solved using the formula

$$x = \frac{-b + \sqrt{b^2 - 4ac}}{2a}$$

NAILIT!

Be careful with the signs when entering numbers into the formula. If you end up with the square root of a negative number you have made a mistake.

[Total: 5 marks]

(2) The right-angled triangle shown has an area of $40\,cm^2$. All measurements are in centimetres. (5 marks, ★★★★)

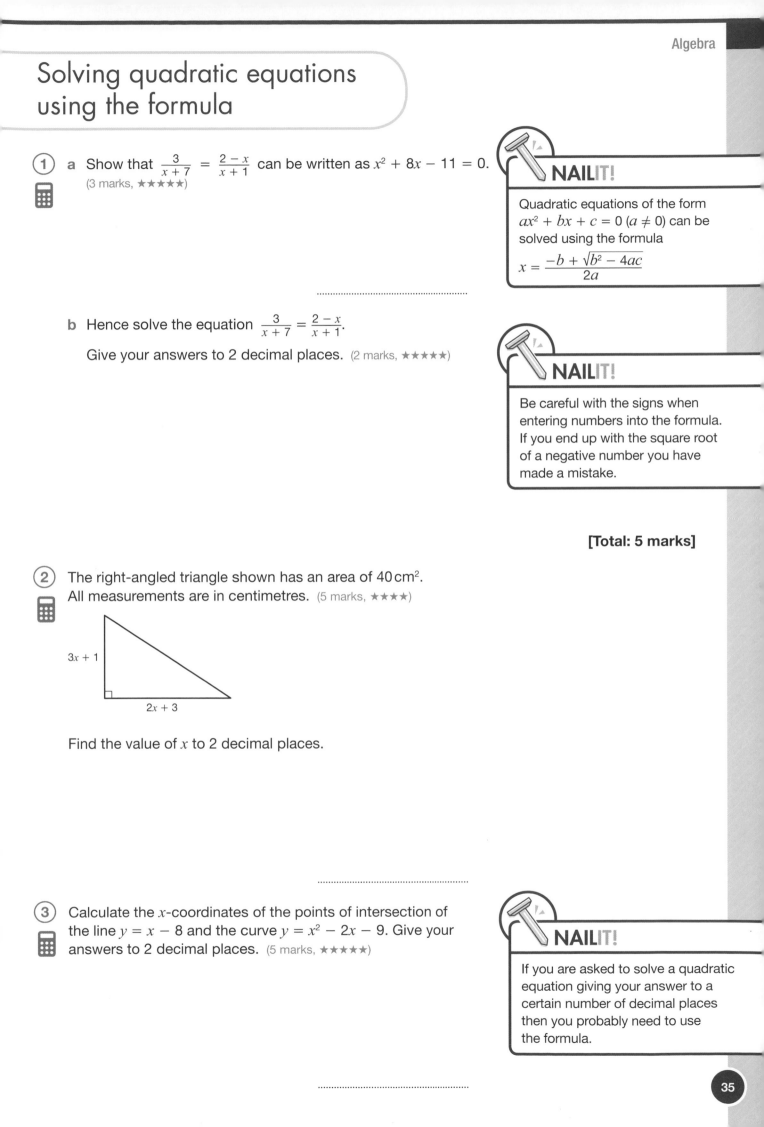

$3x + 1$

$2x + 3$

Find the value of x to 2 decimal places.

....

(3) Calculate the x-coordinates of the points of intersection of the line $y = x - 8$ and the curve $y = x^2 - 2x - 9$. Give your answers to 2 decimal places. (5 marks, ★★★★★)

NAILIT!

If you are asked to solve a quadratic equation giving your answer to a certain number of decimal places then you probably need to use the formula.

....

Solving simultaneous equations

(1) Solve these simultaneous equations algebraically.

$2x - 3y = -5$

$5x + 2y = 16$ (3 marks, ★★★)

The opposite signs for the y terms mean that it is easier to make these terms the same in value but opposite in sign and add the two equations.

NAILIT!

You can solve simultaneous equations:
- graphically, by finding where the equations intersect on a graph
- by eliminating one of the unknowns by adding or subtracting the equations
- by substituting the expression for one variable into the other equation.

NAILIT!

Be careful with the signs when solving simultaneous equations using the elimination method, especially if you have to subtract one equation from the other.

$x =$ .... $y =$ ....

(2) a Plot the graph of $y = 3x - 2$ on the set of axes given. (2 marks, ★★)

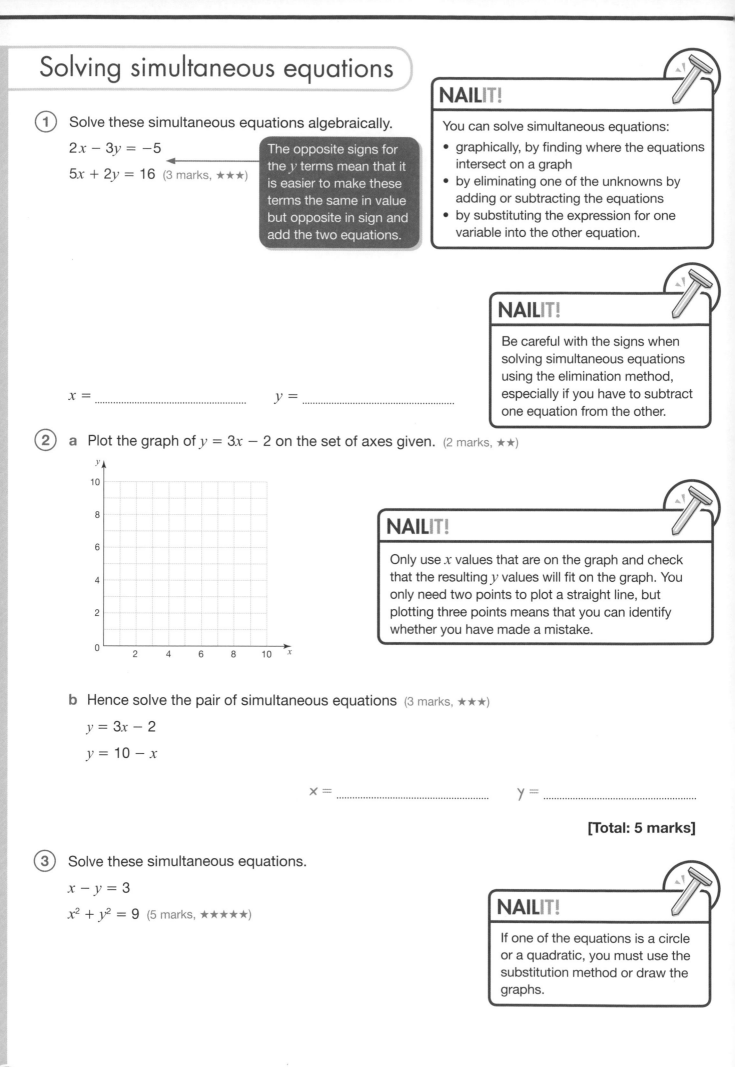

NAILIT!

Only use x values that are on the graph and check that the resulting y values will fit on the graph. You only need two points to plot a straight line, but plotting three points means that you can identify whether you have made a mistake.

b Hence solve the pair of simultaneous equations (3 marks, ★★★)

$y = 3x - 2$

$y = 10 - x$

$x =$ .... $y =$ ....

[Total: 5 marks]

(3) Solve these simultaneous equations.

$x - y = 3$

$x^2 + y^2 = 9$ (5 marks, ★★★★★)

NAILIT!

If one of the equations is a circle or a quadratic, you must use the substitution method or draw the graphs.

$x =$ .... $y =$ ....

Solving inequalities

① Solve these inequalities. (★★★)

a $\dfrac{x + 5}{4} \geq -1$

(1 mark)

b $3x - 4 > 4x + 8$

(2 marks)

[Total: 3 marks]

② Show the inequality $-3 < x \leq 2$ on the number line below. (2 marks, ★★)

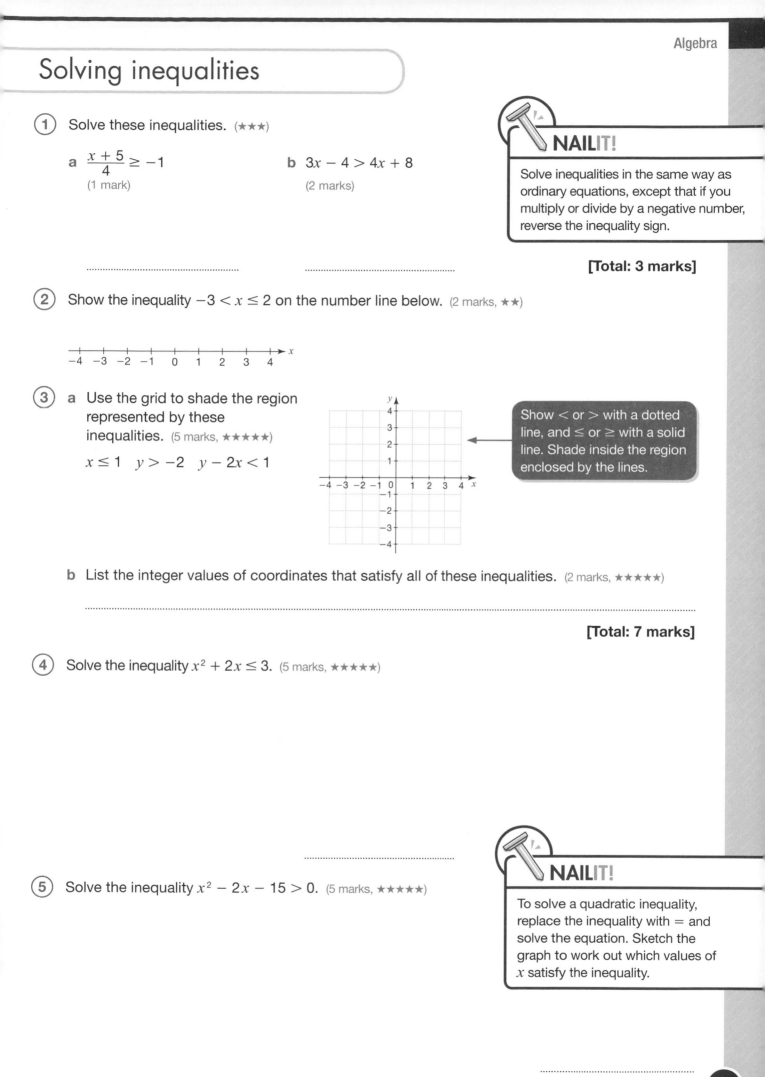

③ a Use the grid to shade the region represented by these inequalities. (5 marks, ★★★★★)

$x \leq 1 \quad y > -2 \quad y - 2x < 1$

b List the integer values of coordinates that satisfy all of these inequalities. (2 marks, ★★★★★)

[Total: 7 marks]

④ Solve the inequality $x^2 + 2x \leq 3$. (5 marks, ★★★★★)

⑤ Solve the inequality $x^2 - 2x - 15 > 0$. (5 marks, ★★★★★)

Problem solving using algebra

1. The length of a rectangular patio is 1 m more than its width.

   The perimeter of the patio is 26 m. Find the area of the patio. (3 marks, ★★★)

   ....

2. The cost of 2 adults' tickets and 5 children's tickets at a circus is £35.

   The cost of 3 adults' tickets and 4 children's tickets is £38.50.

   Find the cost of each type of ticket. (3 marks, ★★★★)

   ....

3. Rachel and Hannah are sisters.

   The product of their ages is 63. In two years' time, the product of their ages will be 99. (★★★★★)

   a Find the sum of their ages. (3 marks)

   ....

   b Rachel is 2 years older than Hannah. How old is Rachel? (2 marks)

   ....

   [Total: 5 marks]

Use of functions

NAILIT!

If $f(x) = x^2 - 1$,
to find $f(1)$ substitute $x = 1$:
$f(1) = 1^2 - 1 = 0$.

(1) $f(x) = 5x + 4$ (★★)

    a Find $f(3)$. (1 mark)

....

STRETCHIT!

If $f(x) = x^2 = y$, can you
express x as a function
of y, $f(y)$?

    b Find the value of x for which $f(x) = -1$. (2 marks)

....

[Total: 3 marks]

(2) If $f(x) = x^2$ and $g(x) = x - 6$, find (★★★)

Apply the function nearest
the x first and then apply the
second function to the answer
$fg(x) = f(g(x))$.

    a $fg(x)$ (1 mark)   **b** $gf(x)$ (2 marks)

....

[Total: 3 marks]

(3) $f(x) = \sqrt{x + 4}$, where $x > -4$

$g(x) = 2x^2 - 3$ for all values of x (★★★★)

    a Find $f(5)$. (1 mark)   **b** Find an expression for $gf(x)$.
                                  Simplify your answer. (2 marks)

....

[Total: 3 marks]

(4) $f(x) = 5x^2 + 3$

Find $f^{-1}(x)$. (3 marks, ★★★★★)

NAILIT!

$f^{-1}(x)$ is the inverse of $f(x)$.

....

Iterative methods

(1) Show that the equation

$$2x^3 - 2x + 1 = 0$$

has a solution between -1 and -1.5. (3 marks, ★★★★)

(2) A sequence is generated using the iterative formula

$$x_{n+1} = x_n^3 + \frac{1}{9}$$

Starting with $x_0 = 0.1$, find x_1, x_2, x_3. For each of these, write down your full calculator display. (3 marks, ★★★★★)

$x_1 =$ ....

$x_2 =$ ....

$x_3 =$ ....

(3) The cubic equation $x^3 - x - 2 = 0$ has a root α between 1 and 2. (★★★★★)

The iterative formula

$$x_{n+1} = (x_n + 2)^{\frac{1}{3}}$$

with $x_0 = 1.5$, can be used to find α.

a Calculate x_4. Give your answer to 3 decimal places. (4 marks)

$x_4 =$ ....

b Prove that this value is also the value of α correct to 3 decimal places. (2 marks)

[Total: 6 marks]

④ **a i** Sketch the graphs of $y = x^3$ and $y = 3 - x$. (2 marks, ★★★★★)

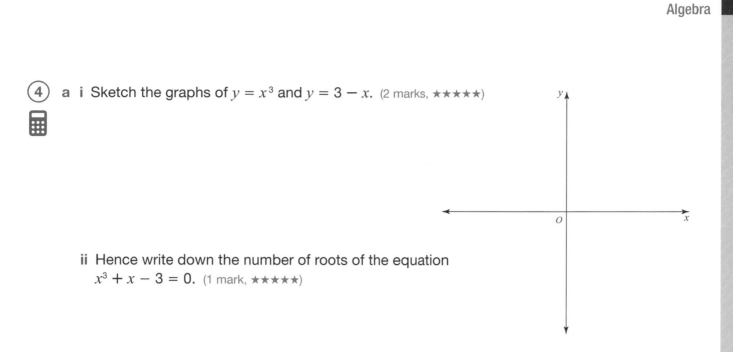

ii Hence write down the number of roots of the equation
$x^3 + x - 3 = 0$. (1 mark, ★★★★★)

....

b The cubic equation $x^3 + x - 3 = 0$ has a root α between 1 and 2.

The iterative formula
$$x_{n+1} = (3 - x_n)^{\frac{1}{3}}$$
with $x_0 = 1.2$, can be used to find α. Calculate x_6. Give your answer to 4 decimal places. (3 marks)

$x_6 =$ ....

[Total: 6 marks]

Equation of a straight line

(1) One of these graphs has the equation $y = 3 - 2x$.

State the letter of the correct graph. (1 mark, ★)

....

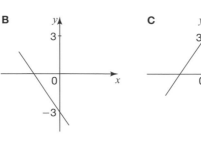

A

B

C

D

(2) **a** The line AB cuts the x-axis at 3 and the y-axis at 4.

Find the gradient of line AB. (2 marks, ★)

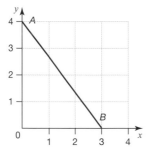

> **NAILIT!**
>
> The gradient of a straight line, $m = \dfrac{\text{change in } y\text{-values}}{\text{change in } x\text{-values}}$
>
> The gradient of the line joining points (x_1, y_1) and (x_2, y_2) is given by $\dfrac{y_2 - y_1}{x_2 - x_1}$.

....

A straight line passes through the points $C(-3, 5)$ and $D(5, 1)$.

b Find the equation of line CD. (3 marks, ★★★)

> **NAILIT!**
>
> Equations of straight lines are of the form $y = mx + c$, where m is the gradient (i.e. the steepness of the line) and c is the intercept on the y-axis.

> Rearrange the equation so that it is in the form $y = mx + c$. Find the gradient and the intercept on the y-axis.

....

> **NAILIT!**
>
> The equation of a straight line with gradient m and which passes through point (x_1, y_1) is given by $y - y_1 = m(x - x_1)$.

c The midpoint of *CD* is *M*. A straight line is drawn through point *M* which is perpendicular to the line *CD*.

Find the equation of this line. (3 marks, ★★★★)

NAILIT!

When two lines are perpendicular to each other, the product of their gradients is -1.

....

[Total: 8 marks]

3 The gradient of line *OP* is 3. Its length is 12. (4 marks, ★★★★★)

You will need to use Pythagoras' theorem.

Find the coordinates of point *P*.
Give each coordinate to 1 decimal place.

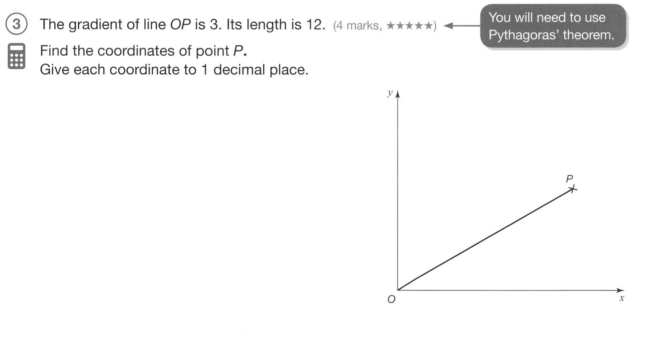

....

Quadratic graphs

1 **a** Using the method of completing the square, solve this quadratic equation.

Give your answers to 1 decimal place.

$x^2 + 4x + 1 = 0$ (4 marks, ★★★★★)

SNAPIT! Quadratic graphs

Quadratic graphs are ∪-shaped if the coefficient of x^2 is positive and ∩-shaped if the coefficient is negative.

....

b Hence sketch the graph of $y = x^2 + 4x + 1$ on the axes provided, including the coordinates of the turning point. (3 marks, ★★★★★)

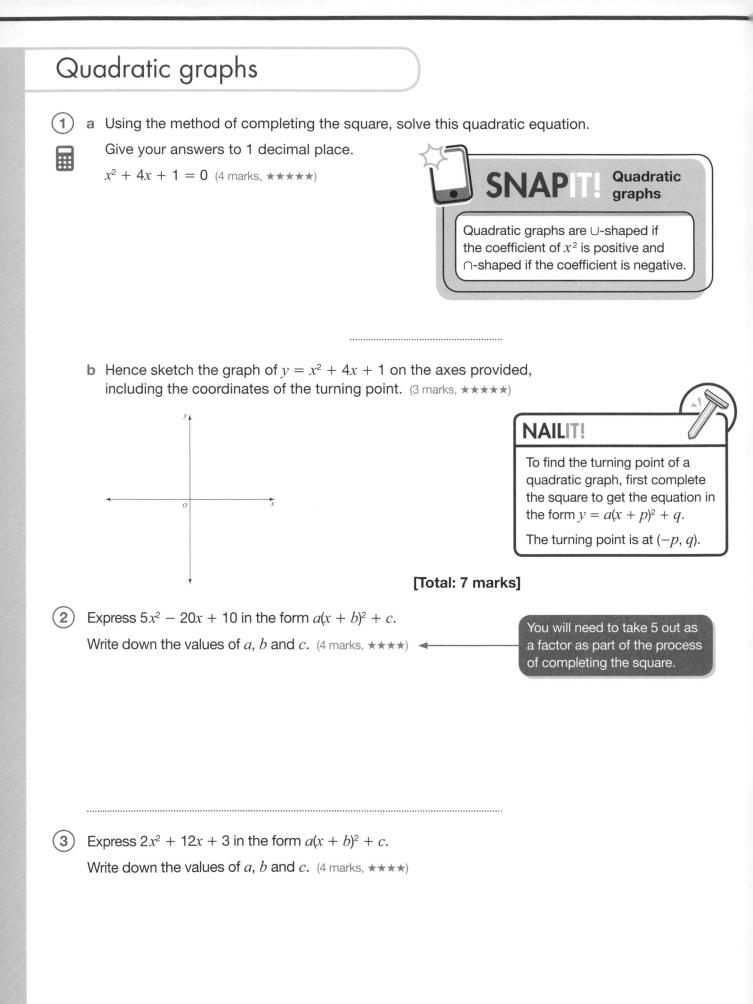

NAILIT!

To find the turning point of a quadratic graph, first complete the square to get the equation in the form $y = a(x + p)^2 + q$.

The turning point is at $(-p, q)$.

[Total: 7 marks]

2 Express $5x^2 - 20x + 10$ in the form $a(x + b)^2 + c$.

Write down the values of a, b and c. (4 marks, ★★★★)

You will need to take 5 out as a factor as part of the process of completing the square.

....

3 Express $2x^2 + 12x + 3$ in the form $a(x + b)^2 + c$.

Write down the values of a, b and c. (4 marks, ★★★★)

....

Recognising and sketching graphs of functions

(1) Fill in the table by inserting the letter of the graph that fits with the equation. (6 marks, ★★★★)

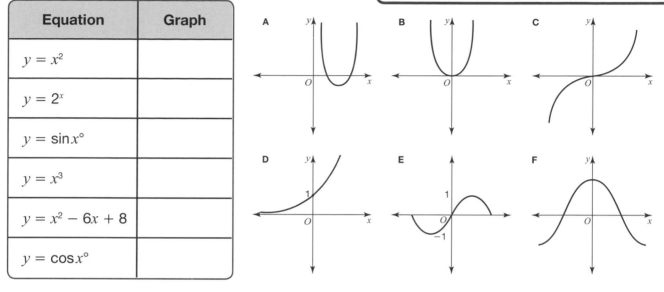

Equation	Graph
$y = x^2$	
$y = 2^x$	
$y = \sin x°$	
$y = x^3$	
$y = x^2 - 6x + 8$	
$y = \cos x°$	

(2) **a** On the axes provided, sketch the graph of $y = \sin x$ for $0° \leq x \leq 360°$. (3 marks, ★★★★)

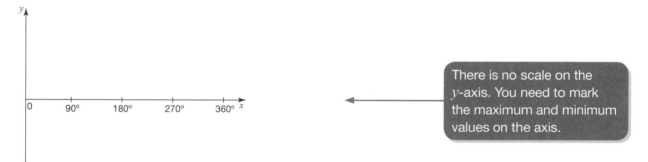

There is no scale on the y-axis. You need to mark the maximum and minimum values on the axis.

b On the axes provided, sketch the graph of $y = \tan x$ for $0° \leq x \leq 360°$. (3 marks, ★★★★)

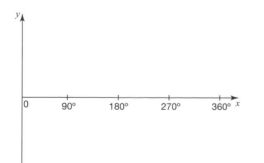

[Total: 6 marks]

(3) Find all the values of θ in the range $0° \leq \theta \leq 360°$ that satisfy

$3 \cos \theta = 1$.

Give your answers to 1 decimal place. (3 marks, ★★★★★)

Translations and reflections of functions

① The quadratic curve $y = f(x)$ passes through the origin and the point (4, 0), and has a turning point at (2, −4). On the same axes, sketch the graphs of (★★★★)

a $y = -f(x)$ (2 marks)

b $y = f(x - 2)$ (2 marks)

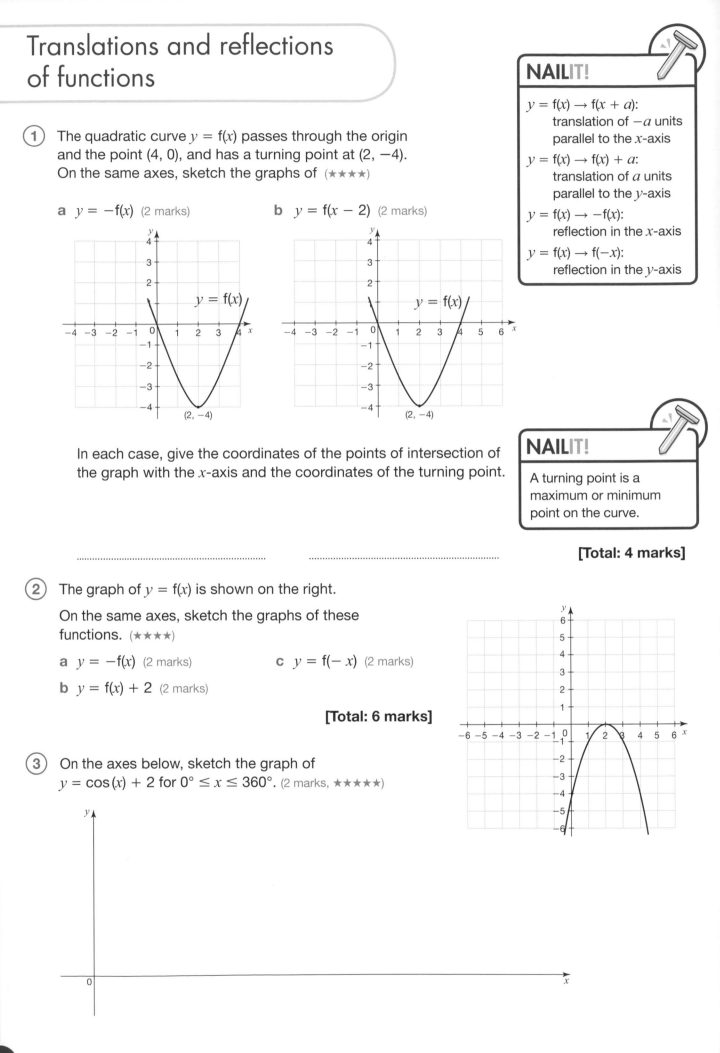

In each case, give the coordinates of the points of intersection of the graph with the x-axis and the coordinates of the turning point.

[Total: 4 marks]

② The graph of $y = f(x)$ is shown on the right.

On the same axes, sketch the graphs of these functions. (★★★★)

a $y = -f(x)$ (2 marks)

b $y = f(x) + 2$ (2 marks)

c $y = f(-x)$ (2 marks)

[Total: 6 marks]

③ On the axes below, sketch the graph of $y = \cos(x) + 2$ for $0° \leq x \leq 360°$. (2 marks, ★★★★★)

Equation of a circle and tangent to a circle

1 Write down the radius of each of these circles. (★★★)

a $x^2 + y^2 = 25$ (1 mark)

b $x^2 + y^2 - 49 = 0$ (1 mark)

c $4x^2 + 4y^2 = 16$ (2 marks)

[Total: 4 marks]

2 A circle has the equation $x^2 + y^2 = 21$. Determine whether the point (4, 3) lies inside or outside this circle. (4 marks, ★★★★★)

> Compare the length of line joining the origin to the point (4, 3) with the radius of the circle.

3 The point $P(5, 7)$ lies on a circle with centre the origin. (★★★★★)

a Find the radius of the circle.
Give your answer as a surd. (2 marks)

WORKIT!

The point $P(2, 3)$ lies on a circle with centre the origin, O.

Find the equation of the tangent to the circle at the point.

1 Find the gradient of the radius OP. $\frac{3}{2}$

2 Find the gradient of the tangent at P. $-\frac{2}{3}$

3 Use $y - y_1 = m(x - x_1)$.

$y - 3 = -\frac{2}{3}(x - 2)$ so
$y = -\frac{2}{3}x + \frac{13}{3}$

b Write down the equation of the circle. (1 mark)

c Find the equation of the tangent to the circle at point P. (3 marks)

[Total: 6 marks]

Real-life graphs

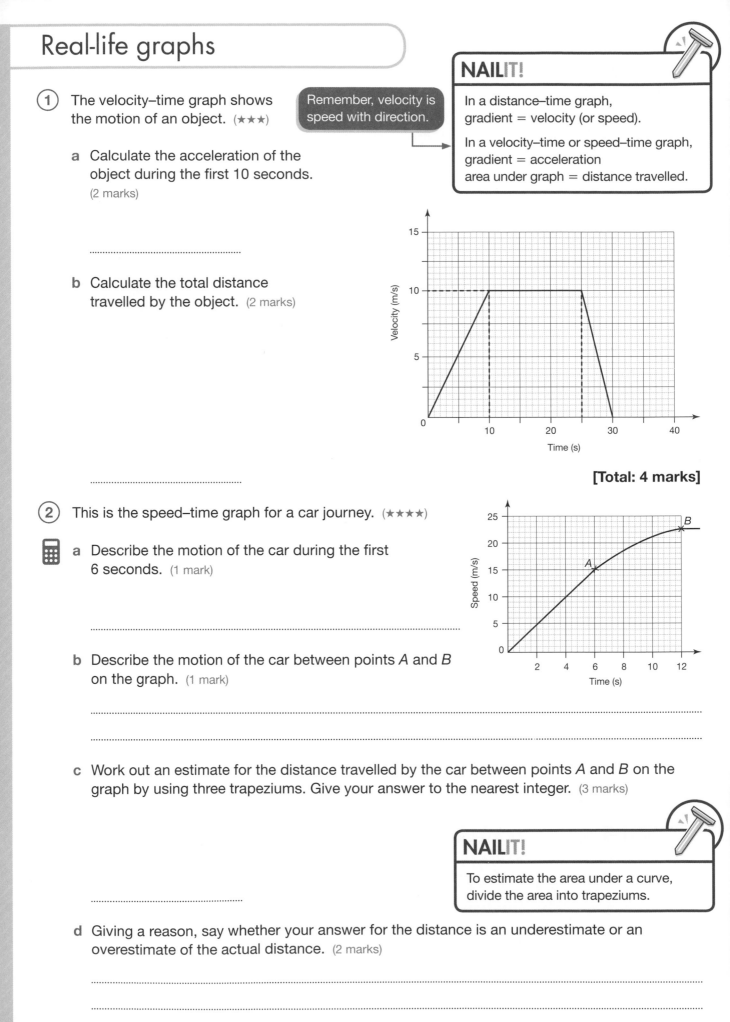

1 The velocity–time graph shows the motion of an object. (★★★)

> Remember, velocity is speed with direction.

a Calculate the acceleration of the object during the first 10 seconds.
(2 marks)

....

b Calculate the total distance travelled by the object. (2 marks)

....

[Total: 4 marks]

2 This is the speed–time graph for a car journey. (★★★★)

a Describe the motion of the car during the first 6 seconds. (1 mark)

....

b Describe the motion of the car between points *A* and *B* on the graph. (1 mark)

....

....

c Work out an estimate for the distance travelled by the car between points *A* and *B* on the graph by using three trapeziums. Give your answer to the nearest integer. (3 marks)

....

d Giving a reason, say whether your answer for the distance is an underestimate or an overestimate of the actual distance. (2 marks)

....

....

[Total: 7 marks]

Generating sequences

1 **a** Work out the next term in each of these sequences. (★★)

   i 16, 8, 4, 2, 1, ... (1 mark)   **iii** 5, 9, 13, 17, ... (1 mark)

....

   ii 3, 9, 27, 81, ... (1 mark)

....

b The following sequence is an arithmetic sequence.

Work out the missing two terms. (★★)

27, ..., ..., −12 (2 marks)

....

[Total: 5 marks]

2 The first two terms of a sequence are

3, 1, ..., ...

The term-to-term rule for this sequence is to multiply the previous term by 2 and subtract 5.

Work out the next two terms of the sequence. (2 marks, ★★)

.... , ...

3 Write down the next two terms for these sequences. (★★★)

a 1, 4, 9, 16, ..., ... (1 mark)

.... , ...

b 1, 3, 6, 10, ..., ... (1 mark)

.... , ...

c 1, 1, 2, 3, 5, ..., ... (1 mark)

.... , ...

[Total: 3 marks]

NAILIT!

You need to be familiar with arithmetic and geometric sequences, as well as the sequences of square numbers, cube numbers, triangular numbers and Fibonacci numbers.

Look for adding or subtracting the same number to get from one term to the next. Then look for multiplying or dividing by the same number.

Work out the difference between 27 and −12. Then divide your answer by three, because three equal distances give two new terms (in the gaps between them).

NAILIT!

A term-to-term rule tells you how to work out the next term in the sequence from the current term.

First work out the difference between the two terms you are given.

NAILIT!

Not all sequences have a constant difference or multiplier. Try squares, cubes, triangular numbers and Fibonacci sequences.

The *n*th term

(1) The first four terms of an arithmetic sequence are

2, 6, 10, 14, ... (★★★)

NAILIT!

You need to be able to write the *n*th term of linear and quadratic sequences as an expression containing *n*.

a Find an expression for the *n*th term of this sequence. (2 marks)

....

b Use your answer to part a to explain why all the terms of this sequence are even. (2 marks)

....

....

c Work out whether 236 is a number in this sequence. (2 marks) ◄—— Write 236 equal to the *n*th term.

....

[Total: 6 marks]

(2) An expression for the *n*th term of a sequence is $9 - n^2$. (★★★)

a Find the 2nd term of this sequence.
(1 mark)

c Explain why 10 cannot be a term in this sequence. (1 mark)

....

....

b Find the 20th term of this sequence. (1 mark)

....

....

[Total: 3 marks]

(3) Find the *n*th term for the sequence

1, 1, 3, 7, 13, ... ◄——

(5 marks, ★★★★★)

There is no constant difference here, so it is likely that this is a quadratic sequence. You need to find the 1st and 2nd differences.

WORKIT!

Find the *n*th term of the sequence   2   3   8   17...

1 Work out the 1st differences.   1   5   9

2 Work out the 2nd differences.   4   4

3 The coefficient of n^2 is $\frac{1}{2}$ × 2nd difference.

*n*th term starts $2n^2$

4 Subtract $2n^2$ from each term.   0   −5   −10   −15

5 Work out the *n*th term for sequence in step 4. $-5n + 5$

6 Combine the two expressions from steps 4 and 5.

*n*th term is $2n^2 - 5n + 5$.

....

Arguments and proofs

(1) Sarah says that all prime numbers are odd. By giving a counter-example, show that her statement is false. (1 mark, ★★)

....

(2) Explain whether each of these statements is true or false. (★★★)

a If n is a positive integer, $2n + 1$ is always greater than or equal to 3. (1 mark)

....

....

b $3(n + 1)$ is always a multiple of 3. (1 mark)

....

....

c If n is a positive integer, $2n - 3$ is always even. (1 mark)

....

....

[Total: 3 marks]

(3) Prove that the sum of any two consecutive integers is always an odd number. (3 marks, ★★★★)

Start by letting the first number be x. Then write an expression in terms of x for the second number.

(4) Prove that $(2x - 1)^2 - (x - 2)^2$ is a multiple of 3 for all integer values of x. (3 marks, ★★★★★)

Multiply out the brackets and then simplify. Look for a factor of 3.

(5) Prove that the difference between the squares of two consecutive odd numbers is always a multiple of 8. (5 marks, ★★★★★)

Ratio
Introduction to ratios

① A bag contains red balls and black balls only. The number of black balls to the number of red balls is in the ratio 3:2.

There are 18 black balls. Work out the number of balls in the bag. (3 marks, ★)

> You know 18 black balls are equal to 3 parts.

NAILIT!

In ratio questions, find the total number of parts by adding the numbers in the ratio together.

② Three daughters are aged 15, 17 and 18 years. They are left £25 000 in a will to be divided between them in the ratio of their ages. Calculate how much each daughter will receive. (3 marks, ★)

WORKIT!

Divide 3.5 kg in the ratio 4:3

1 Work out the total numbers of parts: $3 + 4 = 7$.

2 Work out the value of one part: $3.5 \div 7 = 0.5$ kg.

3 Work out the value of each part of the ratio: 4×0.5 and 3×0.5.

2 kg and 1.5 kg

③ A farm has total area of 800 acres. 40% of the area is devoted to arable crops. The rest is devoted to cattle and sheep in the ratio 9:7.

Work out the land area in acres devoted to sheep. (3 marks, ★★★)

> First work out the area devoted to livestock (60% of 800 acres). Divide your answer in the ratio given.

④ Given that $3x + 1 : x + 4 = 2:3$, find the value of x. (3 marks, ★★★★)

⑤ A wood has pine, oak and ash trees.

The numbers of pine trees and oak trees are in the ratio 5:8.

The numbers of oak trees and ash trees are in the ratio 2:3.

The total number of trees in the wood is 300. Find the number of ash trees. (4 marks, ★★★★)

> Using a multiplier for the second ratio, find a new three-part ratio for all three tree types. Then work out how many parts there are in total.

Scale diagrams and maps

(1) The scale on a map is 1 : 50 000.

Two towns are 10 cm apart on the map.

What is their actual distance apart?
Give your answer in km. (2 marks, ★)

....

(2) A map is drawn to a scale of 1 : 40 000. Find the actual length, in km, of a straight road of length (★)

a 2.3 cm (1 mark)   **b** 3 mm (1 mark) ◀

Multiply each of these lengths by 40 000 and then change the units.

....

[Total: 2 marks]

(3) On a scale drawing of a garden, the length of a path whose actual distance is 40 m is 5 cm. Find the scale of the drawing in the form 1 : n where n is an integer. (3 marks, ★★)

....

(4) The map shows a port and a gas rig.

The actual distance from the port to the gas rig is 12 km.

By taking a measurement from the map, work out the scale of the map.

Give your answer in the form 1 : n where n is an integer. (3 marks, ★★)

....

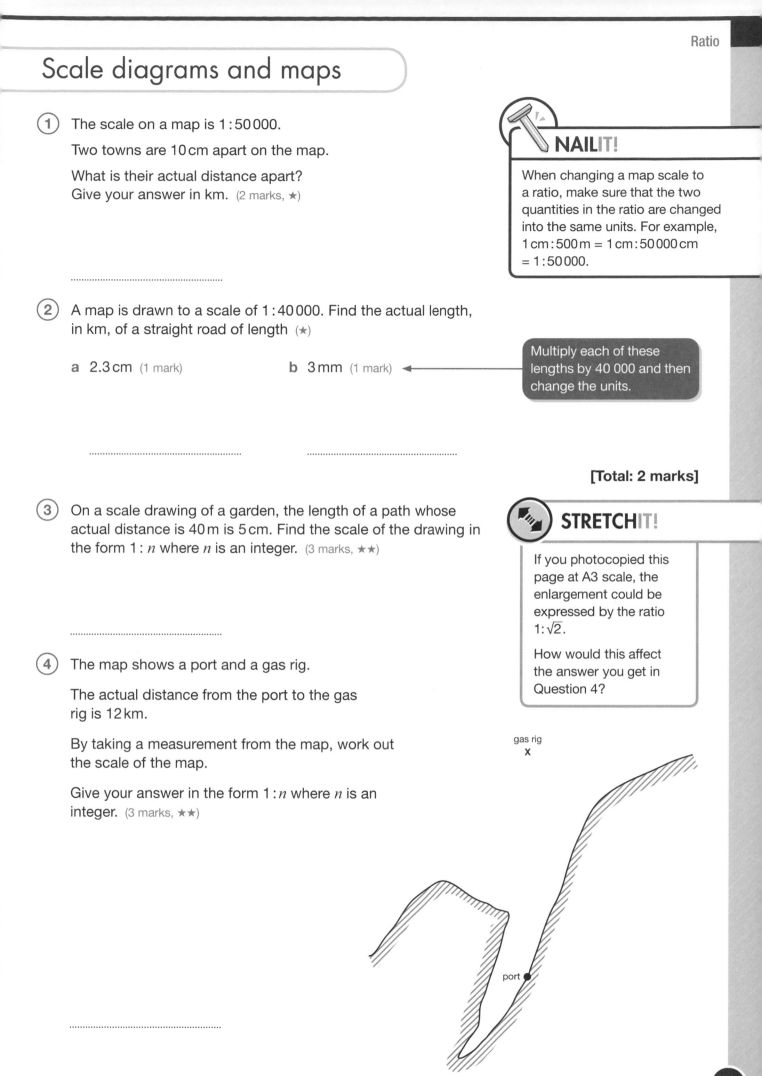

gas rig
X

port ●

Percentage problems

SNAP**IT!** Percentage

Percentage increase/decrease
$= \dfrac{\text{increase/decrease}}{\text{original value}} \times 100\%$

1. A store selling bikes increases the price of a bike from £350 to £385. Find the percentage increase. (2 marks, ★★)

....

2. A football manager initially earns £600 000 per year. On the promotion of his team to the Premier League his earnings increase to £1.1 million per year. Find the percentage increase in his pay to 1 decimal place. (2 marks, ★★)

....

STRETCHIT!

The Whizzie skateboard has been reduced by 20%, from £22 to £17.60. Karl bought his Zoomtown board somewhere else, also for £17.60. He tells Katie, 'At full price, Whizzie skateboards are 20% more expensive than Zoomtown ones'. Is he right?

WORK**IT!**

Work out the multiplier for

Use multipliers to change an amount by a percentage.

a   an increase of 20%

For an increase, multiplier $= 1 + \dfrac{\text{percentage increase}}{100}$

Multiplier $= 1 + \dfrac{20}{100} = 1.2$

b   a decrease of 6%

For a decrease, multiplier $= 1 - \dfrac{\text{percentage increase}}{100}$

Multiplier $= 1 - \dfrac{6}{100} = 0.94$

3. After 3 years a caravan originally costing £25 000 has decreased in value by 28%. What is its new value? (2 marks, ★★)

....

4. The price of a motorbike after a reduction of 12% is £14 300. Find the original price of the motorbike. (3 marks, ★★★)

Start by saying that 88% of the original price is £14 300.

....

5. Fran invests £8000 at an interest rate of 2.8% for 4 years. If simple interest is paid, find the total amount of interest paid over the 4 years. (2 marks, ★★★)

....

Direct and inverse proportion

(1) The pressure of a gas, P, is directly proportional to its temperature, T. (★★★)

    a Write the above statement as an equation. (1 mark)

> **NAILIT!**
>
> When y is directly proportional to x you can write $y = kx$.
>
> When y is inversely proportional to x you can write $y = \frac{k}{x}$

....

    b The pressure is 200 000 Pascals when the temperature is 540 Kelvin.

       Find the value of pressure when the temperature is 200 Kelvin. Give your answer to the nearest whole number. (3 marks)

> Use the pair of values to find the value of the constant.

....   **[Total: 4 marks]**

(2) The cost of building a circular garden pond is directly proportional to the square of the radius of the pond. The cost when the radius is 3 m works out at £480.

    Find the cost of building a circular pond with a radius of 4 m. Give your answer to the nearest whole number. (4 marks, ★★★★)

....

(3) A quantity c is inversely proportional to another quantity h.

    When $c = 3$, $h = 12$ (★★★)

    a write a formula for c in terms of h.
       (1 mark)

    b calculate the value of c when $h = 15$.
       (2 marks)

....

[Total: 3 marks]

(4) Emma goes to France on holiday. (★★★)

a She changes £350 into euros at an exchange rate of £1 = €1.15

Work out how many euros she gets. (1 mark)

....

b When she returns she still has €80.

She changes this back to pounds at an exchange rate of £1 = €1.11.

How many pounds does she get? Give your answer to the nearest penny. (1 mark)

....

c How much would she have saved if she had only changed the money that she needed for the holiday? (2 marks)

....

[Total: 4 marks]

Graphs of direct and inverse proportion and rates of change

1 Which one of these graphs shows that y is directly proportional to x? (1 mark, ★★★)

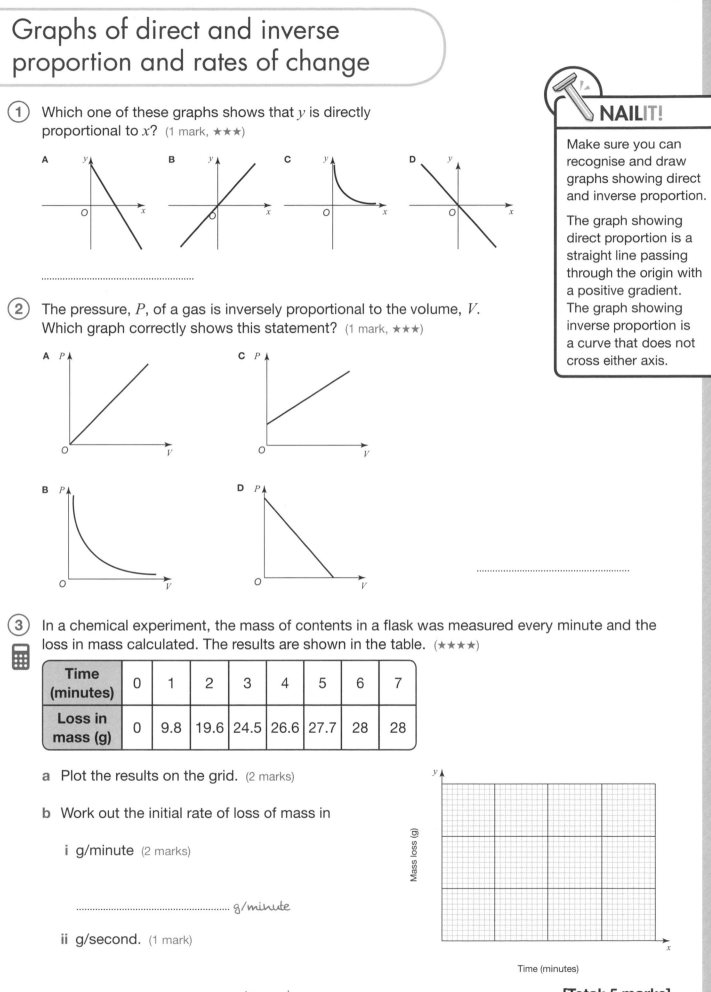

....

2 The pressure, P, of a gas is inversely proportional to the volume, V. Which graph correctly shows this statement? (1 mark, ★★★)

....

3 In a chemical experiment, the mass of contents in a flask was measured every minute and the loss in mass calculated. The results are shown in the table. (★★★★)

Time (minutes)	0	1	2	3	4	5	6	7
Loss in mass (g)	0	9.8	19.6	24.5	26.6	27.7	28	28

a Plot the results on the grid. (2 marks)

b Work out the initial rate of loss of mass in

   i g/minute (2 marks)

.... g/minute

   ii g/second. (1 mark)

.... g/second

[Total: 5 marks]

Growth and decay

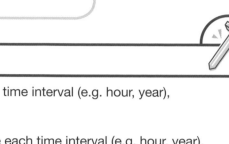
(1) The population of a town is 150 000. Each year the population increases by 6%. (★★★★)

a What will the population be 3 years from now? (2 marks)

> As the population is increasing, the multiplier will be greater than 1.

....

b After how many years will the population have risen to over 200 000? Give your answer to the nearest year. (2 marks)

....

[Total: 4 marks]

(2) Jenny buys an electric car. The car costs £21 000 and it depreciates at a rate of 12% each year. What will be the car's value after 4 years? Give your answer to the nearest whole number. (3 marks, ★★★★)

> Depreciation means that the multiplier will be less than 1.

....

(3) A radioactive isotope halves its activity every 12 seconds.

The initial activity of a sample of the isotope was 100 units.

Find the activity after 2 minutes. Give your answer to 1 significant figure. (4 marks, ★★★★★)

....

Ratios of lengths, areas and volumes

NAILIT!

When using scale factors to work out lengths, areas or volumes, you must make sure that the two shapes are similar.

NAILIT!

If two shapes A and B are similar, we can say l_A is one of the lengths on shape A and l_B is the corresponding length on shape B.

The scale factor for lengths is $\frac{l_B}{l_A}$.

The scale factor for the area A from A to B can be written:

$$\frac{A_B}{A_A} = \left(\frac{l_B}{l_A}\right)^2$$

The scale factor for the volume V from A to B can be written:

$$\frac{V_B}{V_A} = \left(\frac{l_B}{l_A}\right)^3$$

(1) These two triangular prisms are mathematically similar. (★★★★)

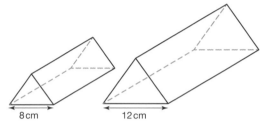

8 cm   12 cm

a Find the scale factor for the volume of the larger prism in relation to the smaller one. (2 marks)

← Work out the scale factor for the lengths first, then use this to find the scale factor for volume.

....

b The area of the triangular cross-section of the small prism is 10 cm². Find the area of cross-section of the large prism. (2 marks)

WORKIT!

Cylinder A and cylinder B are mathematically similar.

The radius of cylinder B is 1.5 times the radius of cylinder A.

The volume of cylinder A is 14 m². Work out the volume of cylinder B.

Scale factor for length = 1.5

Scale factor for volume = 1.5³

Volume of cylinder B = 14 × 1.5³

= 47.25 m³

.... cm²

c The volume of the large prism is 450 cm³. Find the volume of the small prism. Give your answer to the nearest whole number. (2 marks)

.... cm³

[Total: 6 marks]

2 These two solid cuboids are mathematically similar.

The volume of the larger cuboid is 195.3% of the volume of the smaller cuboid.

Calculate the height, h, of the larger cuboid.
Give your answer to the nearest cm. (4 marks, ★★★★★)

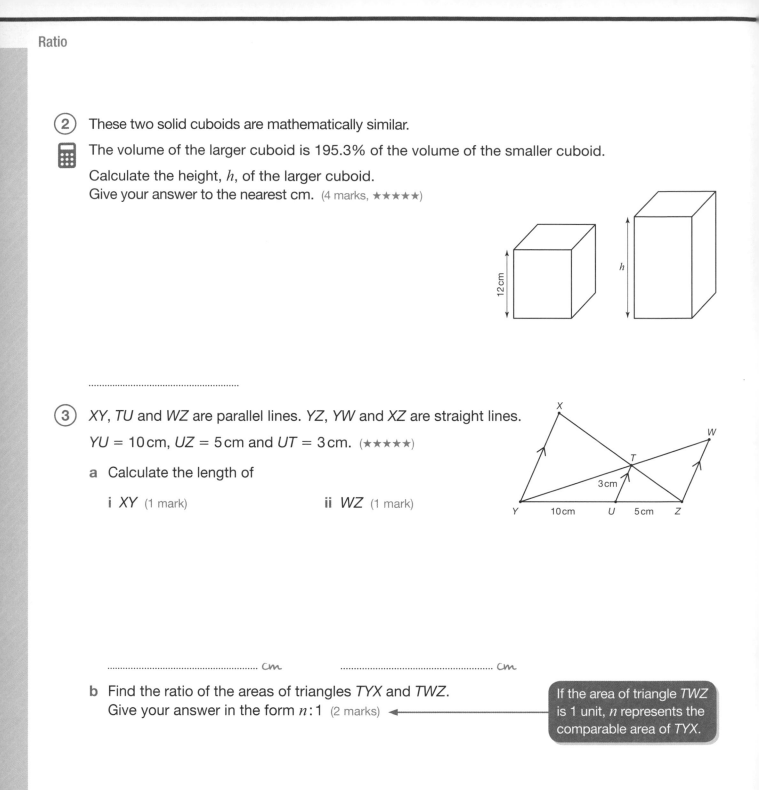

....

3 XY, TU and WZ are parallel lines. YZ, YW and XZ are straight lines.

$YU = 10$ cm, $UZ = 5$ cm and $UT = 3$ cm. (★★★★★)

a Calculate the length of

    i XY (1 mark)   **ii** WZ (1 mark)

.... cm   .... cm

b Find the ratio of the areas of triangles TYX and TWZ.
Give your answer in the form $n:1$ (2 marks) ◄──────

> If the area of triangle TWZ is 1 unit, n represents the comparable area of TYX.

....

[Total: 4 marks]

Gradient of a curve and rate of change

(1) The velocity–time graph shows a car's motion for 90 seconds. (★★★★★)

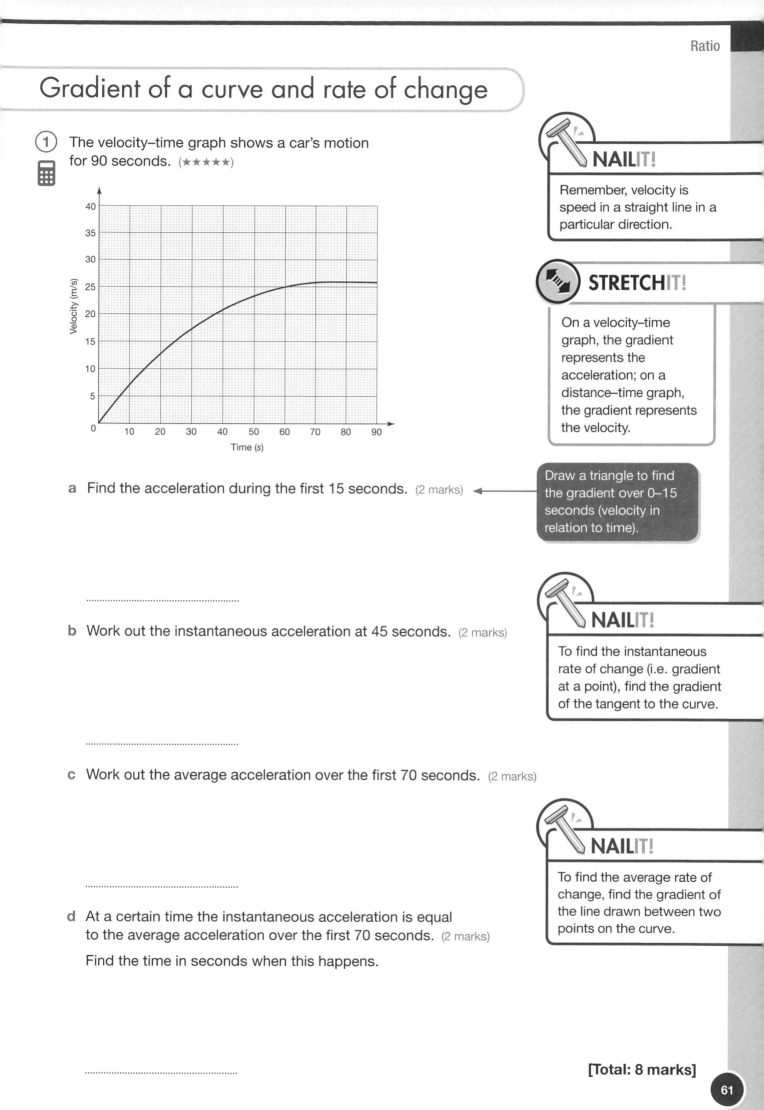

a Find the acceleration during the first 15 seconds. (2 marks)

....

b Work out the instantaneous acceleration at 45 seconds. (2 marks)

....

c Work out the average acceleration over the first 70 seconds. (2 marks)

....

d At a certain time the instantaneous acceleration is equal to the average acceleration over the first 70 seconds. (2 marks)

Find the time in seconds when this happens.

....

[Total: 8 marks]

Converting units of areas and volumes, and compound units

① The formula to work out pressure is pressure = $\frac{\text{force}}{\text{area}}$.

Calculate the pressure produced by a force of 200 N acting on an area of 0.4 m². Give your answer in N/m². (1 mark, ★★)

> The answer requires the force to be in Newtons and the area in m² so no conversions are needed.

....

② Calculate the pressure in Newton/m² if a force of 500 Newtons acts on an area of 200 cm². (2 marks, ★★★)

....

③ The formula for density is density = $\frac{\text{mass}}{\text{volume}}$.

Copper has a density of 8.92 g/cm³.

Find the mass in grams of a length of copper wire with a volume of 12 cm³. Give your answer to the nearest gram. (2 marks, ★★★)

> **NAILIT!**
>
> Compound units are made up of two units: for example m/s for speed, km/h for speed, g/cm³ for density. When entering units into a formula, check whether the values are in the units needed for the final answer. If not, you will need to convert your answer to the correct units.

....

④ A picture has a length of 167 cm and a width of 54 cm.

Joshua wants to work out the area of the picture in m² correct to 2 decimal places.

Here is his working.

Area = length × width = 167 × 54 = 9018 cm²

Area in m² = $\frac{9018}{100}$ = 90.18 m²

Joshua's answer is wrong.

Explain what he has done wrong and work out the correct area in m². (2 marks, ★★★)

....

⑤ Umar drove for 2 hours at 60 km/h. He then drove for 3 hours at 80 km/h.

Work out his average speed for the journey. (3 marks, ★★★)

....

Geometry and measures
2D shapes

NAILIT!

2D shapes can be accurately drawn on a flat sheet of paper.

NAILIT!

You need to remember the names of the regular polygons and also the different types of triangle and quadrilateral, together with their numbers of lines of symmetry and orders of rotational symmetry.

NAILIT!

It is a good idea to do a quick sketch of the shape before deciding on your answer.

(1) State whether each of these statements is true or false. (★)

a The diagonals of a rhombus cut each other at right-angles. (1 mark)

....

b The diagonals of a parallelogram are always the same length. (1 mark)

....

c A kite has one line of symmetry. (1 mark)

....

d A cuboid has 8 vertices and 6 faces. (1 mark)

....

e A trapezium has one pair of parallel sides. (1 mark)

....

f A pentagon has 5 sides of equal length. (1 mark)

....

[Total: 6 marks]

(2) Identify these shapes (★★)

a four equal sides, two pairs of equal angles (1 mark)

....

b two pairs of equal sides, different length diagonals (1 mark)

....

c three sides, order of rotational symmetry 3 (1 mark)

....

d four sides, one line of symmetry, 1 pair of equal angles (1 mark)

....

[Total: 4 marks]

Constructions and loci

① Draw the locus of the point that is always the same distance from points *A* and *B*. (2 marks, ★★)

Construct the perpendicular bisector of *AB*.

A ———————————— *B*

② A goat is secured by a chain of length 1.5 m and a loop on a bar of length 6 m. The loop can travel anywhere along the length of the bar.

6 m

A ●——————————●——————————● *B*

Draw an accurate diagram using a scale of 1 cm to 1 m to show the area in which the goat can graze. Shade the area. (3 marks, ★★★)

③ The diagram below shows two walls, *XY* and *XZ*, at an angle to each other.

A spider waits at point *X*. It then moves along a path such that its distance to lines *XY* and *YZ* is the same. (★★★)

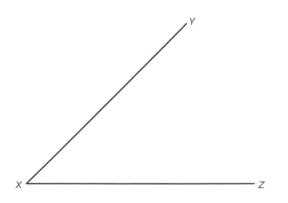

a Using only compasses and a ruler, draw the path of the spider on the above diagram. You should show all your construction lines. (2 marks)

b A fly walks along a path between the two walls so that the distance from point *X* is always 6 cm. Mark the point between the walls where it is possible for the spider's and the fly's paths to cross. (2 marks)

[Total: 4 marks]

Properties of angles

STRETCHIT!

Sometimes shapes in exam questions may be deliberately drawn slightly inaccurately ('not drawn to scale'). For example, they may look like a regular shape when they are not. The diagram below looks like a rhombus, but you would have to prove it using the information in the question and diagram.

NAILIT!

Look for:

- alternate or corresponding angles in parallel lines
- angles in a triangle, including isosceles and equilateral triangles
- angles on a straight line.

① *ABCD* is a rhombus. Point *O* is the point where the diagonals intersect. (★★)

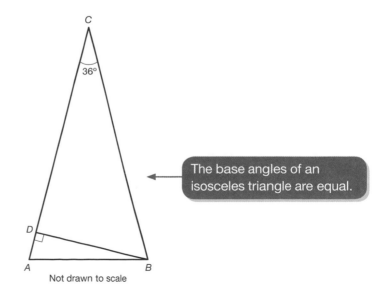

a State, giving a reason, the size of angle *ACB*. (1 mark)

....

....

b State, giving a reason, the size of angle *AOB*. (1 mark)

....

....

Think about the properties of a rhombus.

c Work out the size of angle *BDC*. You must show your workings. (2 marks)

.... °

[Total: 4 marks]

② *ABC* is an isosceles triangle with *AC* = *BC* and angle *ACB* = 36°. *BD* is perpendicular to *AC*.

Find the size of angle *ABD*. Give reasons for each stage of your working. (2 marks, ★★)

The base angles of an isosceles triangle are equal.

.... °

3 *AB*, *CD* and *XY* are straight lines. (★★★)

X

A ———————— 2x° ⟍ 4x° ———————— B

                3x+30°
C ———————————————————— D

Not drawn to scale   Y

a Find the value of x. You must show all your workings. (2 marks)

.... °

b Prove that lines *AB* and *CD* are parallel. (2 marks)

[Total: 4 marks]

4 A needlework group is designing a patchwork quilt. The diagram shows part of their design – a square aligned with a regular pentagon.

Calculate the size of angle x. (4 marks, ★★★)

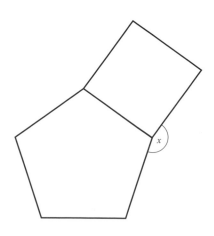

.... °

Congruent triangles

(1) *ABCD* is a parallelogram.

Prove that angle *BAD* = angle *BCD*. (3 marks, ★★★)

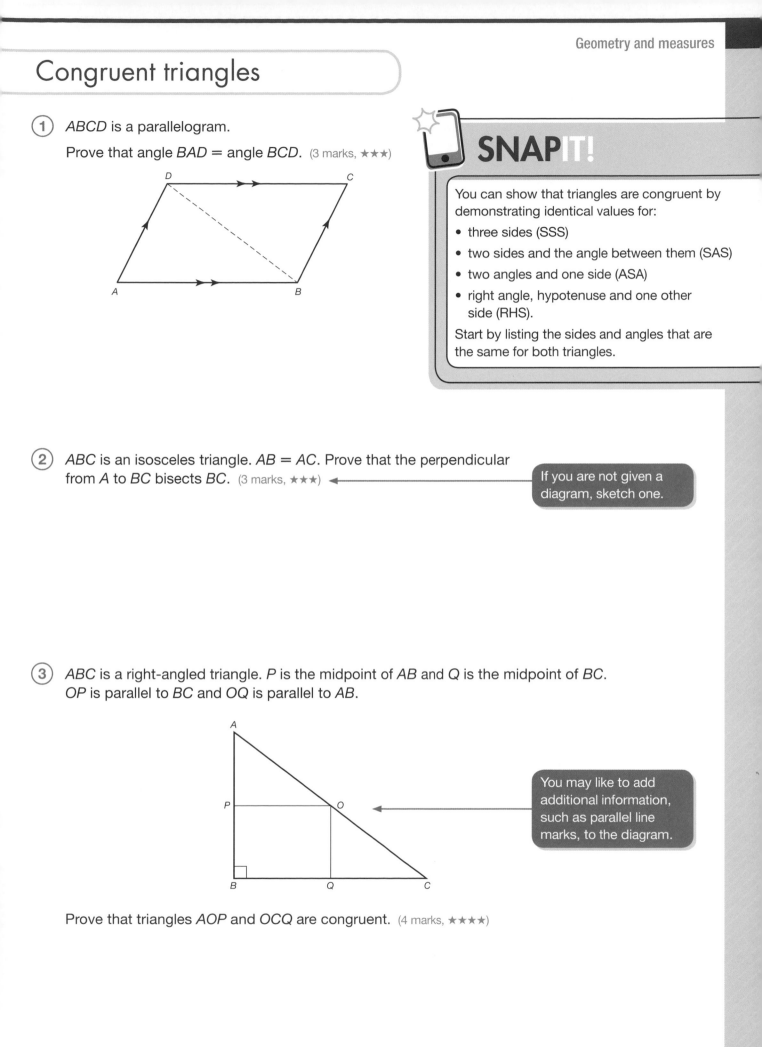

SNAPIT!

You can show that triangles are congruent by demonstrating identical values for:

- three sides (SSS)
- two sides and the angle between them (SAS)
- two angles and one side (ASA)
- right angle, hypotenuse and one other side (RHS).

Start by listing the sides and angles that are the same for both triangles.

(2) *ABC* is an isosceles triangle. *AB* = *AC*. Prove that the perpendicular from *A* to *BC* bisects *BC*. (3 marks, ★★★)

> If you are not given a diagram, sketch one.

(3) *ABC* is a right-angled triangle. *P* is the midpoint of *AB* and *Q* is the midpoint of *BC*. *OP* is parallel to *BC* and *OQ* is parallel to *AB*.

> You may like to add additional information, such as parallel line marks, to the diagram.

Prove that triangles *AOP* and *OCQ* are congruent. (4 marks, ★★★★)

Transformations

① Describe the single transformation that maps triangle A onto triangle B. (1 mark, ★)

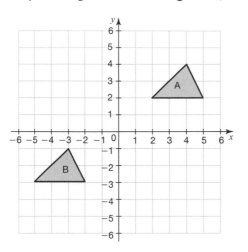

---

② On the grid below, enlarge the triangle by a scale factor of −3 using centre (−3, 3). (3 marks, ★★★★)

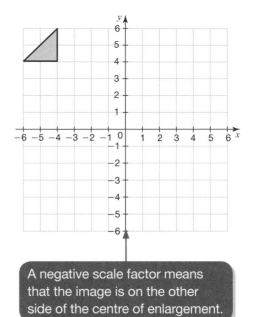

A negative scale factor means that the image is on the other side of the centre of enlargement.

③ Triangles A, B, C and D are is shown on the following grid. (★)

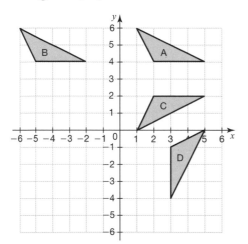

a Describe the single transformation that maps triangle A onto triangle B. (1 mark)

b Describe the single transformation that maps triangle A onto triangle C. (1 mark)

c Describe the single transformation that maps triangle A onto triangle D. (2 marks)

[Total: 4 marks]

Invariance and combined transformations

1 a Triangle A is reflected in the line $x = 0$.
      How many invariant points are there
      on the triangle? (1 mark, ★★★)

....

   b i Triangle A is rotated 90° anticlockwise
         about the origin. The resulting triangle is
         then translated by $\begin{pmatrix} 6 \\ 0 \end{pmatrix}$.

         Draw this new triangle and label it B.
         There is one invariant point on triangle B.
         Give the coordinates of this point.
         (2 marks, ★★★)

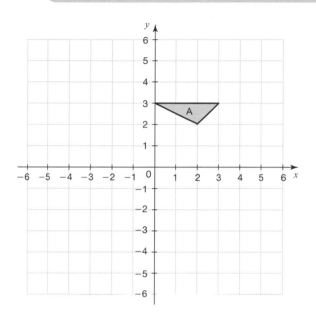

....

      ii Describe the single transformation that
          maps triangle A to triangle B. (1 mark, ★★★)

....

[Total: 4 marks]

2 a Triangle PQR is reflected in the line $x = 4$
      and then rotated 90° anticlockwise about
      point R. Draw the image after these two
      transformations on the grid. (2 marks, ★★★★)

Use tracing paper to work out the
position of a shape after a rotation.

   b Give the coordinates of the invariant
      point after the two transformations.
      (1 mark, ★★★★)

....

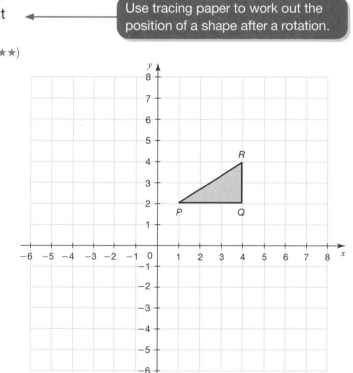

[Total: 3 marks]

3D shapes

(1) Some 3D shapes are drawn below.

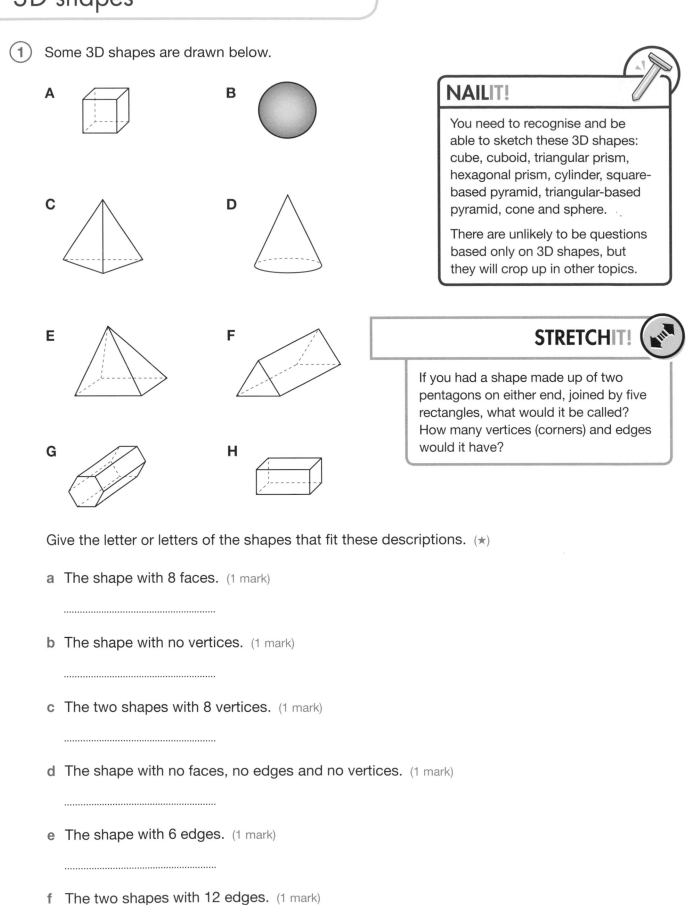

A

B

C

D

E

F

G

H

NAILIT!

You need to recognise and be able to sketch these 3D shapes: cube, cuboid, triangular prism, hexagonal prism, cylinder, square-based pyramid, triangular-based pyramid, cone and sphere.

There are unlikely to be questions based only on 3D shapes, but they will crop up in other topics.

STRETCHIT!

If you had a shape made up of two pentagons on either end, joined by five rectangles, what would it be called? How many vertices (corners) and edges would it have?

Give the letter or letters of the shapes that fit these descriptions. (★)

a The shape with 8 faces. (1 mark)

....

b The shape with no vertices. (1 mark)

....

c The two shapes with 8 vertices. (1 mark)

....

d The shape with no faces, no edges and no vertices. (1 mark)

....

e The shape with 6 edges. (1 mark)

....

f The two shapes with 12 edges. (1 mark)

....

[Total: 6 marks]

Parts of a circle

(1) The diagram shows a circle with centre O.

Name the lines labelled a to d. (★)

a ...Radius... (1 mark)

b ...deameter... (1 mark)

c ...Chord... (1 mark)

d ...Arc... (1 mark)

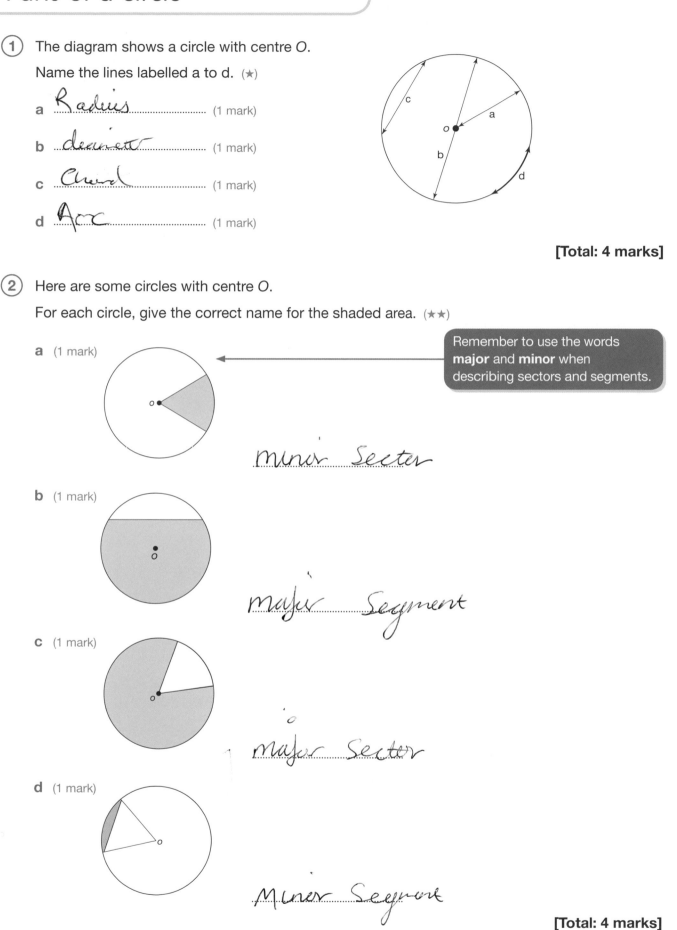

[Total: 4 marks]

(2) Here are some circles with centre O.

For each circle, give the correct name for the shaded area. (★★)

> Remember to use the words **major** and **minor** when describing sectors and segments.

a (1 mark)

minor Sector

b (1 mark)

major Segment

c (1 mark)

major Sector

d (1 mark)

Minor Segment

[Total: 4 marks]

Circle theorems

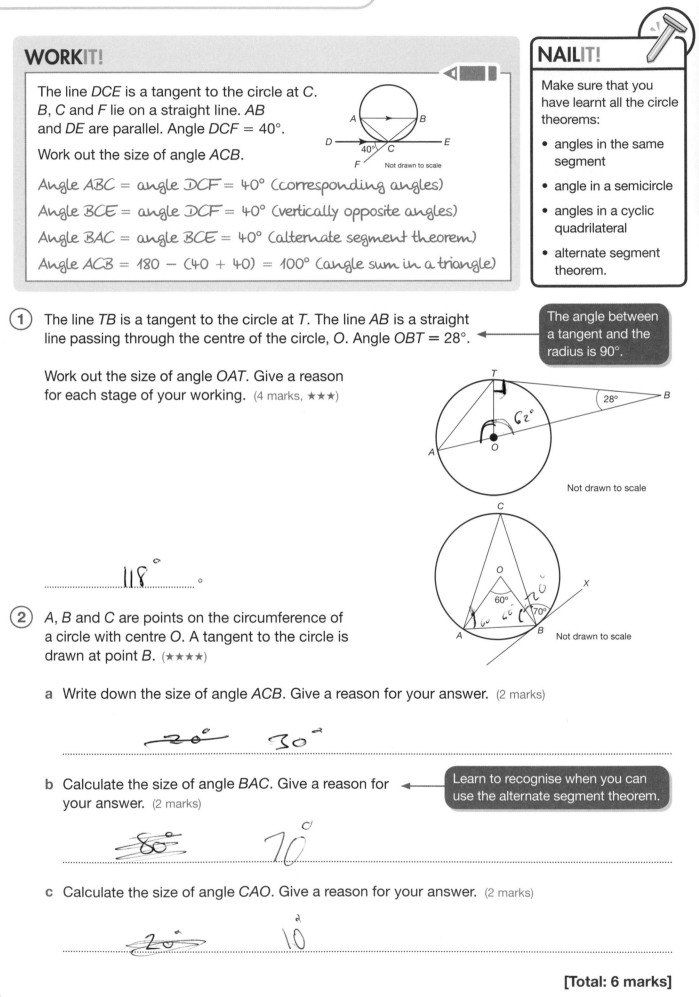
(1) The line *TB* is a tangent to the circle at *T*. The line *AB* is a straight line passing through the centre of the circle, *O*. Angle *OBT* = 28°.

The angle between a tangent and the radius is 90°.

Work out the size of angle *OAT*. Give a reason for each stage of your working. (4 marks, ★★★)

.... 118° ....°

(2) *A*, *B* and *C* are points on the circumference of a circle with centre *O*. A tangent to the circle is drawn at point *B*. (★★★★)

a Write down the size of angle *ACB*. Give a reason for your answer. (2 marks)

.... 20°   30° ....

b Calculate the size of angle *BAC*. Give a reason for your answer. (2 marks)

Learn to recognise when you can use the alternate segment theorem.

.... 80°   70° ....

c Calculate the size of angle *CAO*. Give a reason for your answer. (2 marks)

.... 20°   10 ....

[Total: 6 marks]

Projections

1. This solid shape is made out of seven one-centimetre cubes.

   On the one-centimetre grid shown below, draw a plan view of this solid. (2 marks, ★★)

   > You need to draw in all the edges where there is a change of level.

2. This solid shape is made out of one-centimetre cubes. (★★)

   a On the grids below draw the side elevation from L and the side elevation from R. (2 marks)

   L   R

   b Draw the plan view of the solid on the grid below. (1 mark)

   [Total: 3 marks]

3. The diagram shows the plan, front elevation and side elevation of a solid shape on a centimetre grid.

   Draw a sketch of the solid shape. Write the dimensions on your sketch. (3 marks, ★★★)

   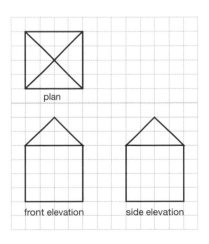

   plan

   front elevation   side elevation

Bearings

WORKIT!

The bearing of *B* from *A* is 285°.
What is the bearing of *A* from *B*?

1 Draw a sketch with the north arrows through both *A* and *B*.
2 Work out the angle between the north arrow and line *AB* at *A*.
3 Use alternate angles to work out the angle between the north arrow and line *AB* at *B*.
4 Use this angle to work out the bearing.

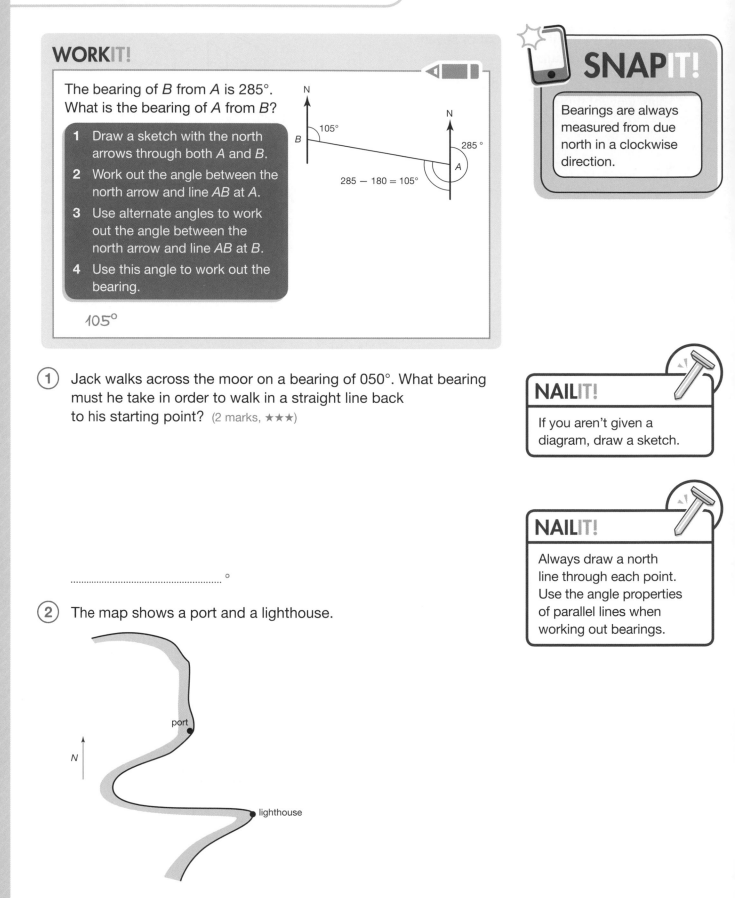

$285 - 180 = 105°$

105°

---

① Jack walks across the moor on a bearing of 050°. What bearing must he take in order to walk in a straight line back to his starting point? (2 marks, ★★★)

.... °

② The map shows a port and a lighthouse.

An offshore wind turbine is on a bearing of 100° from the port and on a bearing of 040° from the lighthouse. Mark this information on the diagram, using construction lines to show the position of the wind turbine. Mark this with an X. (3 marks, ★★★)

Pythagoras' theorem

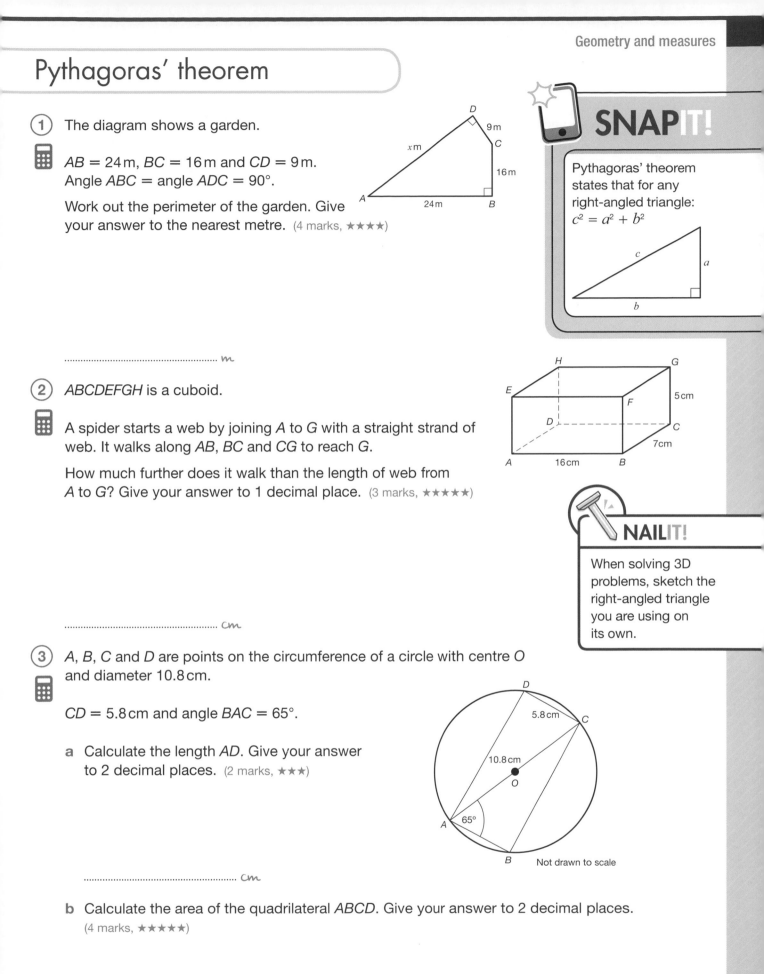

(1) The diagram shows a garden.

$AB = 24$ m, $BC = 16$ m and $CD = 9$ m.
Angle ABC = angle $ADC = 90°$.

Work out the perimeter of the garden. Give your answer to the nearest metre. (4 marks, ★★★★)

.... m

SNAPIT!

Pythagoras' theorem states that for any right-angled triangle:
$c^2 = a^2 + b^2$

(2) *ABCDEFGH* is a cuboid.

A spider starts a web by joining *A* to *G* with a straight strand of web. It walks along *AB*, *BC* and *CG* to reach *G*.

How much further does it walk than the length of web from *A* to *G*? Give your answer to 1 decimal place. (3 marks, ★★★★★)

.... cm

NAILIT!

When solving 3D problems, sketch the right-angled triangle you are using on its own.

(3) *A*, *B*, *C* and *D* are points on the circumference of a circle with centre *O* and diameter 10.8 cm.

$CD = 5.8$ cm and angle $BAC = 65°$.

a Calculate the length *AD*. Give your answer to 2 decimal places. (2 marks, ★★★)

.... cm

b Calculate the area of the quadrilateral *ABCD*. Give your answer to 2 decimal places.
(4 marks, ★★★★★)

.... cm²

[Total: 6 marks]

Area of 2D shapes

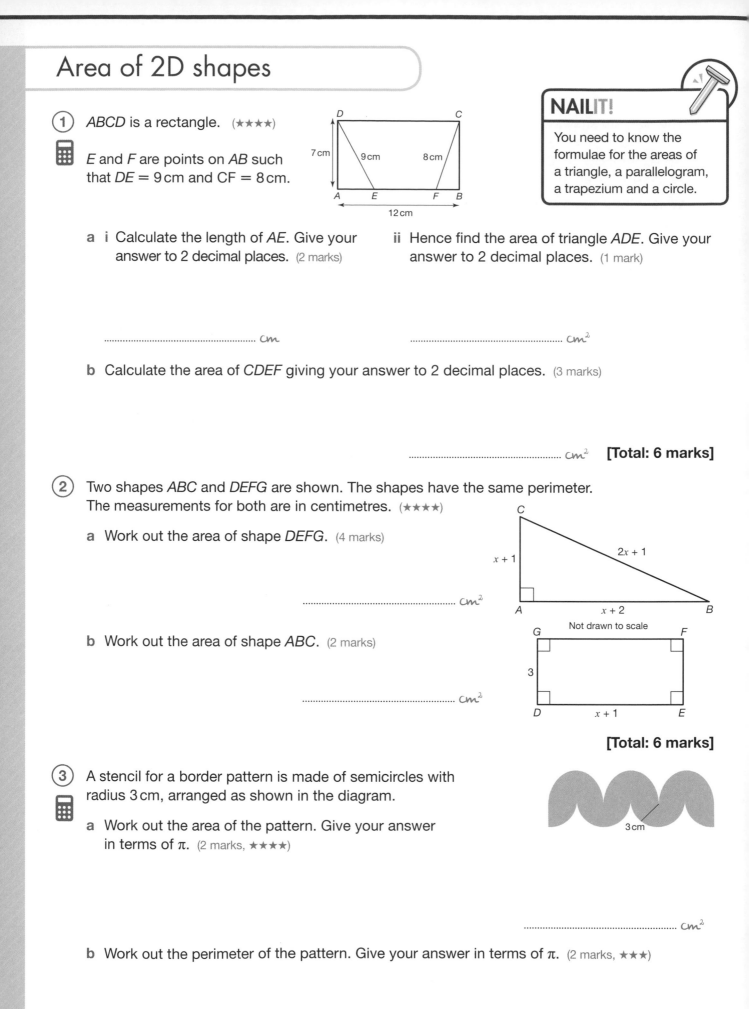

NAIL IT!

You need to know the formulae for the areas of a triangle, a parallelogram, a trapezium and a circle.

(1) *ABCD* is a rectangle. (★★★★)

E and F are points on *AB* such that *DE* = 9 cm and CF = 8 cm.

a i Calculate the length of *AE*. Give your answer to 2 decimal places. (2 marks)

ii Hence find the area of triangle *ADE*. Give your answer to 2 decimal places. (1 mark)

.... cm

.... cm²

b Calculate the area of *CDEF* giving your answer to 2 decimal places. (3 marks)

.... cm² **[Total: 6 marks]**

(2) Two shapes *ABC* and *DEFG* are shown. The shapes have the same perimeter. The measurements for both are in centimetres. (★★★★)

a Work out the area of shape *DEFG*. (4 marks)

.... cm²

b Work out the area of shape *ABC*. (2 marks)

.... cm²

[Total: 6 marks]

(3) A stencil for a border pattern is made of semicircles with radius 3 cm, arranged as shown in the diagram.

a Work out the area of the pattern. Give your answer in terms of π. (2 marks, ★★★★)

.... cm²

b Work out the perimeter of the pattern. Give your answer in terms of π. (2 marks, ★★★)

.... cm²

[Total: 4 marks]

Volume and surface area of 3D shapes

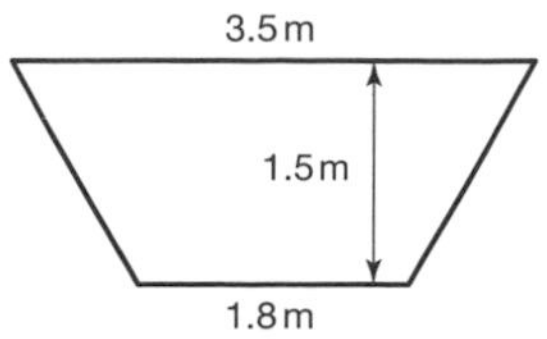

3.5 m

1.5 m

1.8 m

(1) The cross-section of a skip is in the shape of a trapezium. (★★★)

a Calculate the cross-sectional area of the skip shown in the above diagram.
(2 marks)

b The skip is a prism with the cross-sectional area shown and a length of 1.6 m. Calculate the volume of the skip. (2 marks)

> The volume of a prism is the area of cross-section × length.

.... m²

.... m³

[Total: 4 marks]

(2) A glass is in the shape of a hemisphere of radius 4 cm.

How many of these glasses can be filled from a 750 ml bottle? (5 marks, ★★★★★)

....

(3) **a** A solid cone has base radius 4.8 cm and perpendicular height 5.6 cm.

Work out the slant height of the cone. Give your answer to 1 decimal place. (2 marks, ★★★★)

> Slant height is the distance from the tip to the circular edge. Use Pythagoras' theorem to calculate it.

.... cm

b A sphere has the same total surface area as the total surface area of the cone.

Work out the radius of the sphere. Give your answer to 1 decimal place. (4 marks, ★★★★★)

SNAPIT!

Surface area of sphere $= 4\pi r^2$
Curved surface area of cone $= \pi r l$
Volume of a cone $= \frac{1}{3}\pi r^2 h$
Volume of a cylinder $= \pi r^2 h$

.... cm **[Total: 6 marks]**

(4) The diagram shows a cone with a radius of 2 cm and height 12 cm. The cone is filled to the top with water. All of the water is then poured into a cylinder with radius 5 cm.

Work out the depth of the water in the cylinder. (4 marks, ★★★★)

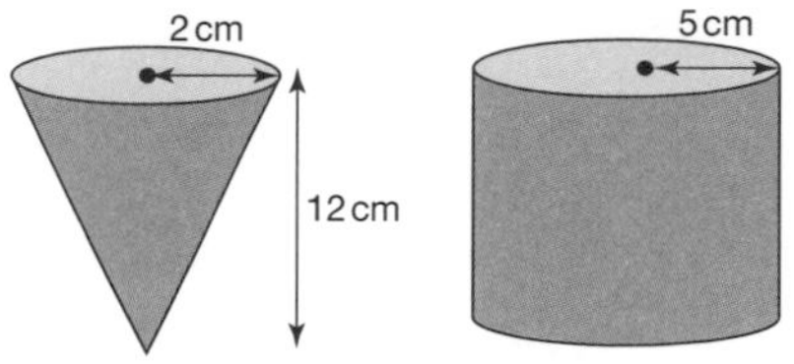

2 cm

5 cm

12 cm

.... cm

Trigonometric ratios

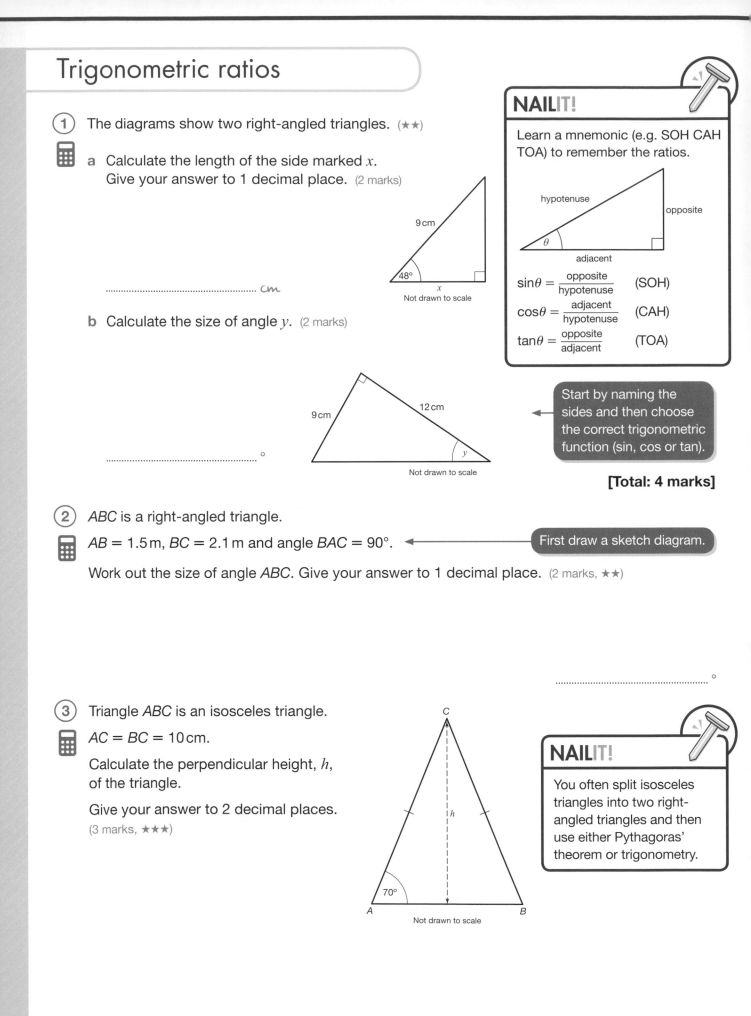

(1) The diagrams show two right-angled triangles. (★★)

a Calculate the length of the side marked x.
Give your answer to 1 decimal place. (2 marks)

.... cm

b Calculate the size of angle y. (2 marks)

9 cm

48°

x

Not drawn to scale

9 cm

12 cm

y

Not drawn to scale

.... °

NAILIT!

Learn a mnemonic (e.g. SOH CAH TOA) to remember the ratios.

hypotenuse

opposite

θ

adjacent

$\sin\theta = \dfrac{\text{opposite}}{\text{hypotenuse}}$ (SOH)

$\cos\theta = \dfrac{\text{adjacent}}{\text{hypotenuse}}$ (CAH)

$\tan\theta = \dfrac{\text{opposite}}{\text{adjacent}}$ (TOA)

Start by naming the sides and then choose the correct trigonometric function (sin, cos or tan).

[Total: 4 marks]

(2) *ABC* is a right-angled triangle.

$AB = 1.5$ m, $BC = 2.1$ m and angle $BAC = 90°$. ◄———— First draw a sketch diagram.

Work out the size of angle *ABC*. Give your answer to 1 decimal place. (2 marks, ★★)

.... °

(3) Triangle *ABC* is an isosceles triangle.

$AC = BC = 10$ cm.

Calculate the perpendicular height, h, of the triangle.

Give your answer to 2 decimal places.

(3 marks, ★★★)

C

h

70°

A   B

Not drawn to scale

NAILIT!

You often split isosceles triangles into two right-angled triangles and then use either Pythagoras' theorem or trigonometry.

.... cm

(4) The diagram shows a cuboid with length 5.5 cm, height 2.5 cm and depth 3 cm.

Work out the angle between the longest diagonal and the base of the box. (3 marks, ★★★★★)

NAILIT!

In 3D questions, sketch the right-angled triangles that you need. Then use Pythagoras' theorem or trigonometry to work out what you are asked to find.

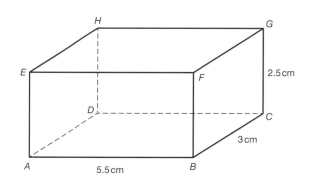

.... °

STRETCHIT!

In the diagram above, *HB* would be an internal diagonal. Can you find the lines that are the other internal diagonals? How many are there in all?

HB is in a right-angled triangle, *HDB*. Write down the right-angled triangles for the other diagonals that you found.

Exact values of sin, cos and tan

(1) Show that

$$\tan 45° + \cos 60° = \frac{3}{2} \quad \text{(3 marks, ★★★★)}$$

(2) a Write the exact value of (★★★)

   i $\sin 45°$ (1 mark)

> Draw a sketch if you need to. When asked for the exact value, use surds.

....

   ii $\cos 45°$ (1 mark)

....

b Hence prove that $\frac{\sin 45°}{\cos 45°} = \tan 45°$
(3 marks, ★★★★)

[Total: 5 marks]

(3) The diagram shows a right-angled triangle *ABC*. (★★★)

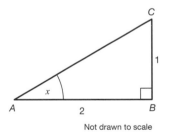

Not drawn to scale

a Find the length of *AC*. Give your answer as a surd. (2 marks)

....

b Write down the exact values of

   i $\sin x$ (1 mark)

....

   ii $\cos x$ (1 mark)

....

c Hence show that $(\sin x)^2 + (\cos x)^2 = 1$
(3 marks)

[Total: 7 marks]

(4) Without using a calculator, show that $\tan 30° + \tan 60° + \cos 30° = \frac{11\sqrt{3}}{6}$ (3 marks, ★★★★★)

Sectors of circles

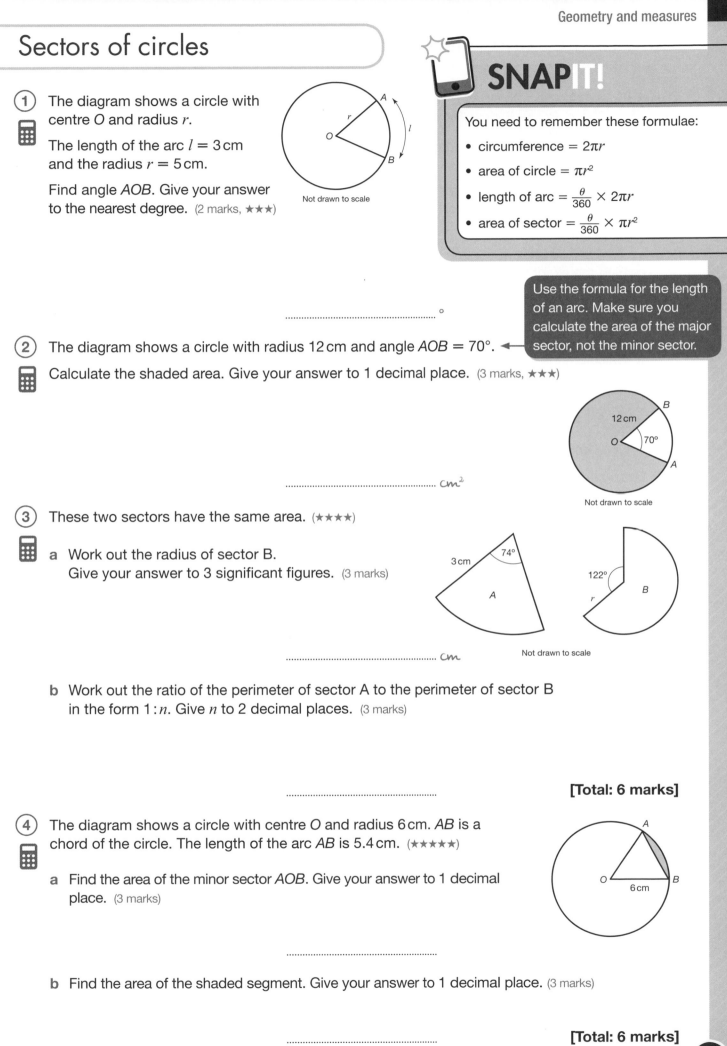

(1) The diagram shows a circle with centre O and radius r.

The length of the arc $l = 3$ cm and the radius $r = 5$ cm.

Find angle AOB. Give your answer to the nearest degree. (2 marks, ★★★)

Not drawn to scale

.... °

(2) The diagram shows a circle with radius 12 cm and angle $AOB = 70°$. ← Use the formula for the length of an arc. Make sure you calculate the area of the major sector, not the minor sector.

Calculate the shaded area. Give your answer to 1 decimal place. (3 marks, ★★★)

.... cm^2

Not drawn to scale

(3) These two sectors have the same area. (★★★★)

a Work out the radius of sector B.
Give your answer to 3 significant figures. (3 marks)

3 cm   74°   122°   B   r

A

.... cm

Not drawn to scale

b Work out the ratio of the perimeter of sector A to the perimeter of sector B in the form $1 : n$. Give n to 2 decimal places. (3 marks)

....

[Total: 6 marks]

(4) The diagram shows a circle with centre O and radius 6 cm. AB is a chord of the circle. The length of the arc AB is 5.4 cm. (★★★★★)

a Find the area of the minor sector AOB. Give your answer to 1 decimal place. (3 marks)

....

b Find the area of the shaded segment. Give your answer to 1 decimal place. (3 marks)

A   O   6 cm   B

....

[Total: 6 marks]

Sine and cosine rules

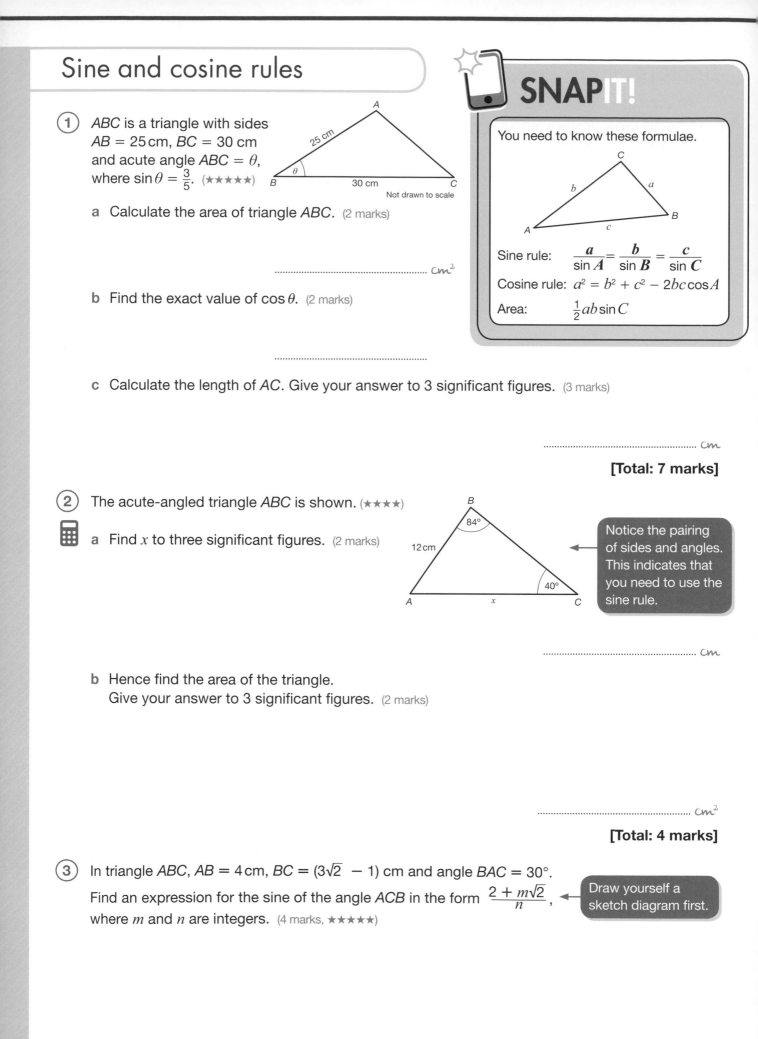

SNAPIT!

You need to know these formulae.

Sine rule: $\dfrac{a}{\sin A} = \dfrac{b}{\sin B} = \dfrac{c}{\sin C}$

Cosine rule: $a^2 = b^2 + c^2 - 2bc\cos A$

Area: $\dfrac{1}{2}ab\sin C$

1 *ABC* is a triangle with sides
AB = 25 cm, *BC* = 30 cm
and acute angle *ABC* = θ,
where $\sin\theta = \dfrac{3}{5}$. (★★★★★)

25 cm

θ

B   30 cm   C

A

Not drawn to scale

a Calculate the area of triangle *ABC*. (2 marks)

.... cm²

b Find the exact value of $\cos\theta$. (2 marks)

....

c Calculate the length of *AC*. Give your answer to 3 significant figures. (3 marks)

.... cm

[Total: 7 marks]

2 The acute-angled triangle *ABC* is shown. (★★★★)

a Find *x* to three significant figures. (2 marks)

B

84°

12 cm

40°

A   x   C

Notice the pairing
of sides and angles.
This indicates that
you need to use the
sine rule.

.... cm

b Hence find the area of the triangle.
Give your answer to 3 significant figures. (2 marks)

.... cm²

[Total: 4 marks]

3 In triangle *ABC*, *AB* = 4 cm, *BC* = $(3\sqrt{2} - 1)$ cm and angle *BAC* = 30°.

Find an expression for the sine of the angle *ACB* in the form $\dfrac{2 + m\sqrt{2}}{n}$,

where *m* and *n* are integers. (4 marks, ★★★★★)

Draw yourself a
sketch diagram first.

....

Vectors

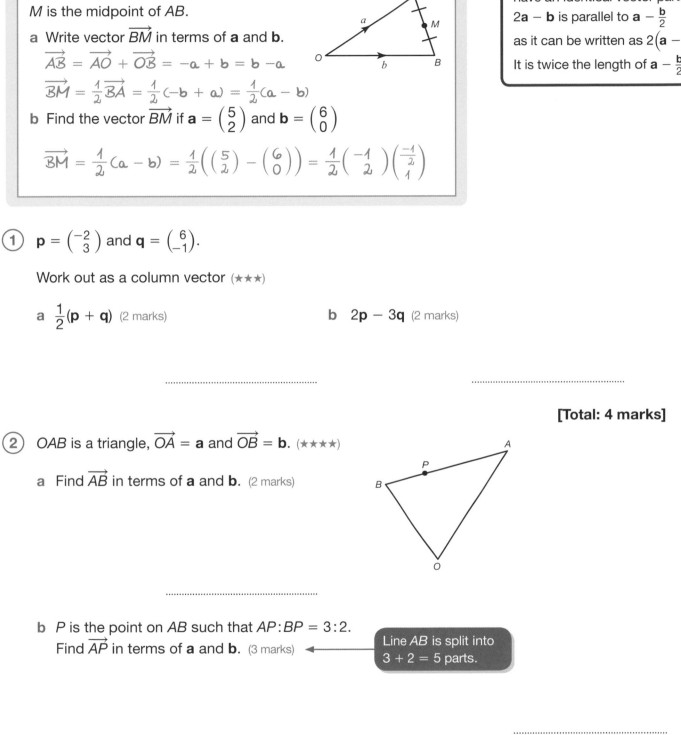

WORKIT!

In triangle OAB, $\overrightarrow{OA} = \mathbf{a}$ and $\overrightarrow{OB} = \mathbf{b}$.

M is the midpoint of AB.

a Write vector $\overrightarrow{BM}$ in terms of **a** and **b**.

$\overrightarrow{AB} = \overrightarrow{AO} + \overrightarrow{OB} = -\mathbf{a} + \mathbf{b} = \mathbf{b} - \mathbf{a}$

$\overrightarrow{BM} = \frac{1}{2}\overrightarrow{BA} = \frac{1}{2}(-\mathbf{b} + \mathbf{a}) = \frac{1}{2}(\mathbf{a} - \mathbf{b})$

b Find the vector $\overrightarrow{BM}$ if $\mathbf{a} = \begin{pmatrix} 5 \\ 2 \end{pmatrix}$ and $\mathbf{b} = \begin{pmatrix} 6 \\ 0 \end{pmatrix}$

$\overrightarrow{BM} = \frac{1}{2}(\mathbf{a} - \mathbf{b}) = \frac{1}{2}\left(\begin{pmatrix} 5 \\ 2 \end{pmatrix} - \begin{pmatrix} 6 \\ 0 \end{pmatrix}\right) = \frac{1}{2}\begin{pmatrix} -1 \\ 2 \end{pmatrix}\begin{pmatrix} \frac{-1}{2} \\ 1 \end{pmatrix}$

NAILIT!

If two vectors are parallel they have an identical vector part:

$2\mathbf{a} - \mathbf{b}$ is parallel to $\mathbf{a} - \frac{\mathbf{b}}{2}$

as it can be written as $2\left(\mathbf{a} - \frac{\mathbf{b}}{2}\right)$.

It is twice the length of $\mathbf{a} - \frac{\mathbf{b}}{2}$.

---

(1) $\mathbf{p} = \begin{pmatrix} -2 \\ 3 \end{pmatrix}$ and $\mathbf{q} = \begin{pmatrix} 6 \\ -1 \end{pmatrix}$.

Work out as a column vector (★★★)

a $\frac{1}{2}(\mathbf{p} + \mathbf{q})$ (2 marks)

b $2\mathbf{p} - 3\mathbf{q}$ (2 marks)

....

....

[Total: 4 marks]

(2) OAB is a triangle, $\overrightarrow{OA} = \mathbf{a}$ and $\overrightarrow{OB} = \mathbf{b}$. (★★★★)

a Find $\overrightarrow{AB}$ in terms of **a** and **b**. (2 marks)

....

b P is the point on AB such that $AP:BP = 3:2$.
Find $\overrightarrow{AP}$ in terms of **a** and **b**. (3 marks) ◄—— Line AB is split into $3 + 2 = 5$ parts.

....

c Q is a point on OA such that $OQ:QA = 2:3$. Are lines QP and OB parallel? Explain your answer. (2 marks)

....

[Total: 7 marks]

(3) *ABC* is a triangle and *ACD* is a straight line.

$\overrightarrow{AC} = $ **a** and $\overrightarrow{AB} = 3$**b**.

P is a point on *AB* such that $AP:PB = 2:1$.

M is the midpoint of *BC*.

C is the midpoint of *AD*. (★★★★★)

a Find $\overrightarrow{BC}$ in terms of **a** and **b**. (1 mark)

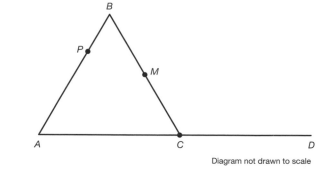

Diagram not drawn to scale

....

b Show that *PMD* is a straight line. (4 marks)

[Total: 5 marks]

Probability
The basics of probability

1 Two dice are thrown and the scores are added together. (★★)

The sample space diagram for the total score is shown below. The first row has been completed.

a Complete the sample space diagram.

(2 marks)

Dice 1

		1	2	3	4	5	6
		1	2	3	4	5	6
Dice 2	1	2	3	4	5	6	7
	2						
	3						
	4						
	5						
	6						

b Find the probability of obtaining a score that is a prime number. (1 mark)

....

c What is the most likely score? Give a reason for your answer. (1 mark)

....

....

[Total: 4 marks]

2 A box of chocolates contains three different types of chocolate.

The table shows the number of chocolates of each type.

	Truffle	Mint	Caramel
Number of chocolates	$2x + 1$	x	$2x$

You need to find the value of x first.

A chocolate is chosen at random. The probability of the chocolate being a mint is $\frac{4}{21}$. (★★★)

a Calculate the total number of chocolates in the box. You must show your working.

(3 marks)

b Calculate the probability of choosing a truffle. (1 mark)

....

....

[Total: 3 marks]

3 Amy and Beth each throw a fair dice.

Calculate the probability that the score on Amy's dice is (★★★★)

a equal to the score on Beth's dice (2 marks)

b greater than the score on Beth's dice. (2 marks)

....

....

[Total: 4 marks]

Probability experiments

50 light bulbs were tested; 44 lasted for 1500 hours or more, and 6 for less than 1500 hours.

In a batch of 1000 light bulbs, how many would be expected to last for less than 1500 hours?

Relative frequency for less than 1500 hours
$= \frac{6}{50}$, so frequency $= \frac{6}{50} \times 1000$
$= 120$ light bulbs

NAILIT!

When probability experiments are conducted (e.g. throwing a dice, spinning a spinner, etc.)

Relative frequency $= \frac{\text{frequency of particular event}}{\text{total trials in experiment}}$

The value for the relative frequency will only approach the theoretical probability over a very large number of trials.

1. A pentagonal unbiased spinner with sides numbered 1 to 5 was spun 100 times. (★★)

The frequency that the spinner landed on each number was recorded in the table below.

Score on spinner	1	2	3	4	5
Frequency	18	23	22	19	18

a Abdul says that the spinner must be biased because if it was fair the frequency for each score would be the same. Explain why Abdul is wrong. (1 mark)

....

....

b Calculate the relative frequency of obtaining a score of 3 on the spinner. Give your answer as a fraction in its simplest form. (2 marks)

....

c If the spinner was spun 500 times, use the relative frequency to estimate how many times the spinner would give a score of 4. (1 mark)

....

[Total: 4 marks]

(2) A hexagonal spinner with sides numbered from 1 to 6 has the following relative frequencies of scores from 1 to 6. (★★★)

Score	1	2	3	4	5	6
Relative frequency	$3x$	0.05	$2x$	0.25	0.2	0.1

a Calculate the value of x. (2 marks)

....

b Calculate the relative frequency of obtaining a score of 1. (1 mark)

....

c The spinner was spun 80 times. Estimate how many times the score was 5. (1 mark)

....

[Total: 4 marks]

STRETCHIT!

Playing cards are interesting because they offer a lot of different probabilities. If you draw a card, it could be: red, black, a heart, a diamond, a club, a spade, an ace, a picture card, a 5 and so on. Think about how these different probabilities affect the games people play – and conjuring tricks too.

The AND and OR rules

(1) A card is chosen at random from a pack of 52 playing cards and noted. The card is returned, the pack shuffled and a second card chosen. (★★★)

Find the probability the two cards were

a two picture cards (1 mark)

....

b an ace and a picture card (1 mark)

....

c the queen of diamonds and the queen of hearts. (1 mark)

....

[Total: 3 marks]

(2) A bag contains 10 marbles: 3 red, 5 blue and 2 green.

A marble is removed from the bag and its colour noted before it is put back into the bag.

Another marble is removed and its colour is also noted. (★★★)

a Explain what is meant by independent events. (1 mark)

Not related

b Calculate the probability that two red marbles were picked. (2 marks)

$9/100$ ✓

c Calculate the probability that the two marbles were red and blue. (2 marks)

$30/100$ ✓

[Total: 5 marks]

(3) The probability that Aisha is given maths homework on a certain day is $\frac{3}{5}$. The probability she is given French homework is $\frac{3}{7}$. The probability she is given geography homework is $\frac{1}{4}$. (★★★★)

a Work out the probability that she is given homework in all three subjects.

(2 marks)

$9/140$ ✓

b Work out the probability that she is not given homework in any of these subjects.

(2 marks)

$24/140$ ✓

[Total: 4 marks]

Tree diagrams

WORKIT!

A box contains 7 counters: 3 red and 4 blue.

A counter taken at random from the box is red.

A second counter is taken at random.

What is the probability that the second counter is red if

a the first counter was put back in the box

Number of red counters in box = 3

Total number of counters = 7

P(red) = $\frac{3}{7}$

b the first counter was not put back in the box?

Number of red counters in box = 2

Total number of counters = 6

P(red) = $\frac{2}{6}$ = $\frac{1}{3}$

NAILIT!

In independent events, the probabilities do not affect each other. In conditional events, the probability of the second outcome depends on what the first outcome was.

(1) From a group of children consisting of 4 girls and 6 boys, 2 children are picked at random to take part in an interview.

Work out the probability that (★★★★)

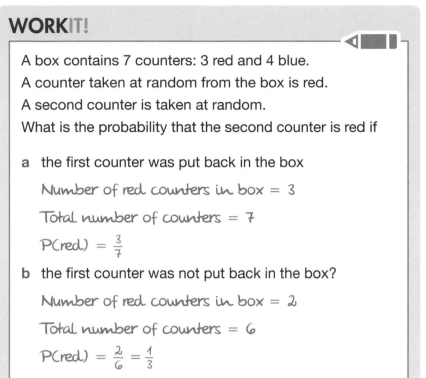

a two girls are chosen (2 marks)

NAILIT!

Draw a tree diagram showing the possible selections. The probability of a particular sequence is found by multiplying the probabilities along the branches making up the sequence.

b a boy and a girl are chosen. (2 marks)

$4\,\frac{8}{90}$

[Total: 4 marks]

(2) There are 450 pupils in a junior school, 56% of whom are boys.

The ratio of boys who have school dinners to the boys who have a packed lunch is 2:1. The ratio of girls who have school dinners to the girls who have a packed lunch is 5:4 (★★★★)

> Make sure you understand the difference between probability trees and frequency trees.

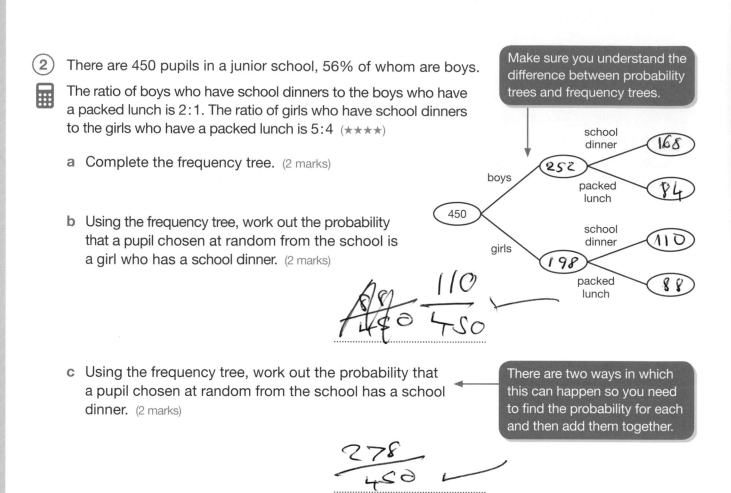

a Complete the frequency tree. (2 marks)

b Using the frequency tree, work out the probability that a pupil chosen at random from the school is a girl who has a school dinner. (2 marks)

$$\frac{88}{450} \frac{110}{450}$$

c Using the frequency tree, work out the probability that a pupil chosen at random from the school has a school dinner. (2 marks)

> There are two ways in which this can happen so you need to find the probability for each and then add them together.

$$\frac{278}{450} \checkmark$$

[Total: 6 marks]

(3) A bag contains 9 marbles, of which 5 are red, 3 are green and 1 is yellow. Three marbles are chosen at random from the bag.

Calculate the probability of choosing the following. Give each answer correct to 3 decimal places. (★★★★★)

> Draw a probability tree to help you.

a 1 marble of each colour (2 marks)

$$\frac{30}{168}$$

b 0 green marbles (2 marks)

$$\frac{.10}{42}$$

c 3 marbles of the same colour (2 marks)

$$\frac{11}{84}$$

[Total: 6 marks]

Venn diagrams and probability

1 The Venn diagram shows the universal set with two subsets A and B. (★★★)

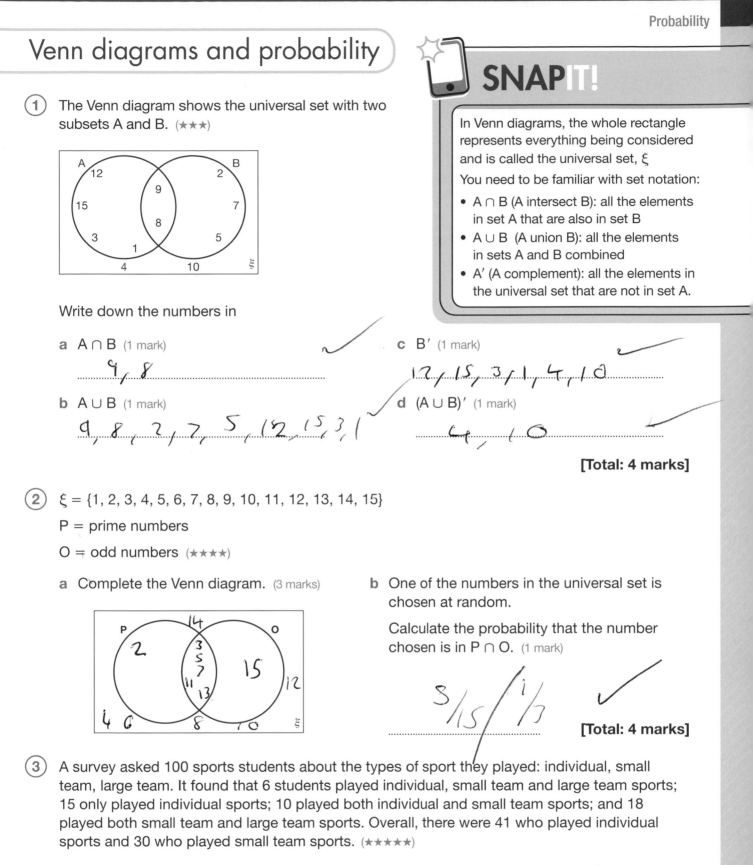

Write down the numbers in

a A ∩ B (1 mark)

9, 8

b A ∪ B (1 mark)

9, 8, 2, 7, 5, 12, 15, 3, 1

c B′ (1 mark)

12, 15, 3, 1, 4, 10

d (A ∪ B)′ (1 mark)

4, 10

[Total: 4 marks]

2 ξ = {1, 2, 3, 4, 5, 6, 7, 8, 9, 10, 11, 12, 13, 14, 15}

P = prime numbers

O = odd numbers (★★★★)

a Complete the Venn diagram. (3 marks)

b One of the numbers in the universal set is chosen at random.

Calculate the probability that the number chosen is in P ∩ O. (1 mark)

$\frac{5}{15} / \frac{1}{3}$ ✓

[Total: 4 marks]

3 A survey asked 100 sports students about the types of sport they played: individual, small team, large team. It found that 6 students played individual, small team and large team sports; 15 only played individual sports; 10 played both individual and small team sports; and 18 played both small team and large team sports. Overall, there were 41 who played individual sports and 30 who played small team sports. (★★★★★)

a A sports student is chosen at random. What is the probability that the student plays team sports? (3 marks)

$\frac{85}{100}$

b A student is chosen at random from those students who play large team sports. What is the probability that they also play small team sports? (2 marks)

$\frac{18}{73}$

[Total: 5 marks]

Statistics
Sampling

1 Dorota and Ling are going to perform a survey at two different schools. Both schools have 900 students. Dorota intends to take a sample of 50 students and Ling intends to take a sample of 100 students. (★★★)

> **NAILIT!**
>
> A sample is a smaller part of the population. It is very important that the sample is not biased in any way and represents the population.

   a Whose survey is likely to be more reliable and why? (2 marks)

....

....

   b There are 500 girls and 400 boys in Dorota's school. How many boys should she include in her sample? (2 marks)

> The proportion of boys in the sample should be the same as the proportion in the school.

....

[Total: 4 marks]

2 A survey is to be taken about parking in a town centre. There are 1000 parking spaces in the town, and a sample of people who use them is to be used for the survey.

State two ways of making sure that the sample is unbiased. (2 marks, ★★)

....

....

....

....

3 A health survey is being carried out to look at peoples' opinions about smoking.

It has been found that 52% of the population is male and that 12% of them smoke. Of the females, 10% smoke.

> You may use a calculator for these questions.

A sample of 400 people is to be used, with the same male-to-female ratio as the overall population. Each person in the sample is to be given a questionnaire to fill in. (★★★)

   a How many questionnaires should be given to females who do not smoke? (2 marks)

....

   b How many questionnaires should be given to males who smoke? (2 marks)

....

[Total: 4 marks]

Two-way tables and pie charts

1. The two-way table gives information about the favourite TV soap of a sample of 100 students in a college. (★★★)

	Coronation Street	EastEnders	Emmerdale	Total
Boys	12	31	20	
Girls		12		37
Total	30			

a Complete the two-way table. (2 marks)

b A student is picked at random from the college. Find the probability that the student's favourite TV soap is Emmerdale. (1 mark)

....

c A girl is picked at random. Find the probability that her favourite TV soap is EastEnders. (1 mark)

....

NAILIT!

In a two-way table, the numbers must add up across the rows and down the columns to give the totals.

[Total: 4 marks]

2. The table gives information about eye colour in a secondary school class.

Complete the pie chart to show this information. (4 marks, ★★)

Colour	Frequency
Blue	11
Green	3
Brown	16

Work out the angle that represents one person. Then find the angle for each of the frequencies in the table.

3. The headteacher of a school recorded the number of students who were late each day over a 21-day period. Here are the results. (★★)

12	8	10	22	20	17	12
5	12	18	11	8	16	4
10	12	11	3	4	9	26

a Find the median number of students late each day during this period. (1 mark)

b Find the range of the number of students who were late. (1 mark)

....

[Total: 2 marks]

Line graphs for time series data

1. The total number of insects found in a sample area of wildflower meadow for the first 6 months of the year are recorded here. (★)

Month	Jan	Feb	Mar	Apr	May	June
Insects	83	89	133	270	445	480

   a Describe the trend in insect numbers over the period for the data. (1 mark)

   ....

   ....

   b A scientist wants to study the diversity of insect life in the meadow. Circle the measure of average that would be most useful in choosing a month to conduct the study. (1 mark)

   Mean   Median   Mode

   [Total: 2 marks]

2. This table records the percentage of people in a social club who said they would like a vegetarian option included on the menu for the annual dinner. (★★)

Year	1996	2000	2004	2008	2012	2016
Percentage	45	49	53	57	67	74

   a Plot the time series graph. Make the x-axis go up to at least 2025. (2 marks)

   b Describe the trend in attitudes towards vegetarian food. (1 mark)

   ....

   ....

   c What percentage of the people are likely to want a vegetarian option to be included on the menu for the 2025 annual dinner? (1 mark)

   ....

   d Explain why your prediction in part c may not be reliable. (1 mark)

   ....

   ....

   [Total: 5 marks]

Averages and spread

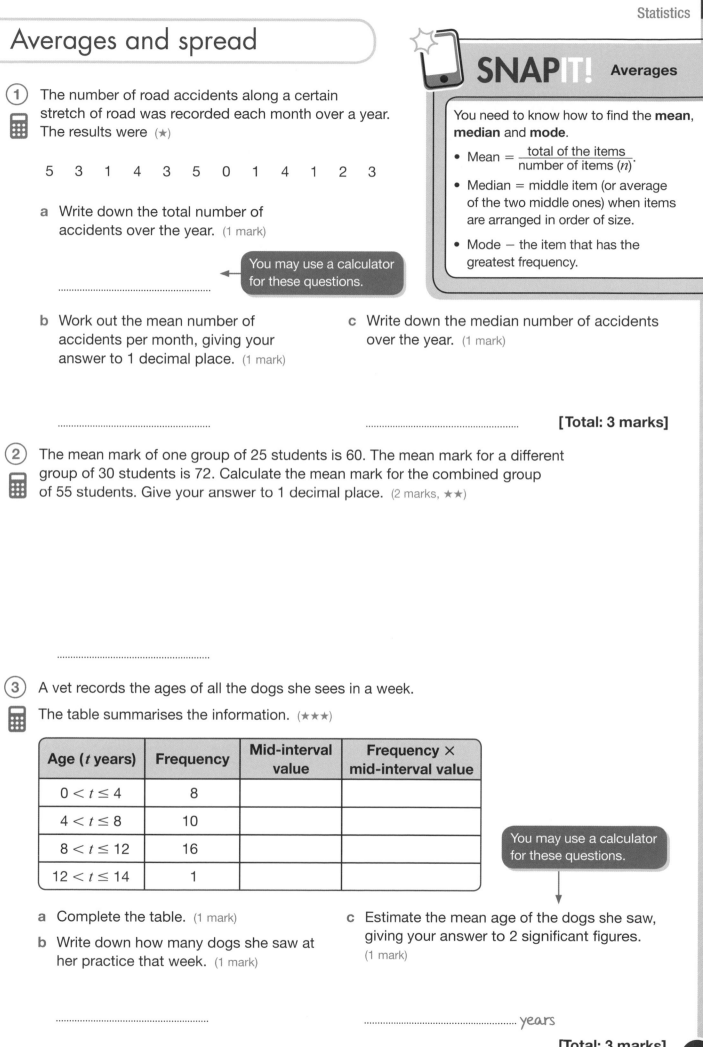

(1) The number of road accidents along a certain stretch of road was recorded each month over a year. The results were (★)

5   3   1   4   3   5   0   1   4   1   2   3

a Write down the total number of accidents over the year. (1 mark)

> You may use a calculator for these questions.

....

b Work out the mean number of accidents per month, giving your answer to 1 decimal place. (1 mark)

c Write down the median number of accidents over the year. (1 mark)

....

....   **[Total: 3 marks]**

(2) The mean mark of one group of 25 students is 60. The mean mark for a different group of 30 students is 72. Calculate the mean mark for the combined group of 55 students. Give your answer to 1 decimal place. (2 marks, ★★)

....

(3) A vet records the ages of all the dogs she sees in a week.

The table summarises the information. (★★★)

Age (t years)	Frequency	Mid-interval value	Frequency × mid-interval value
$0 < t \le 4$	8		
$4 < t \le 8$	10		
$8 < t \le 12$	16		
$12 < t \le 14$	1		

> You may use a calculator for these questions.

a Complete the table. (1 mark)

b Write down how many dogs she saw at her practice that week. (1 mark)

c Estimate the mean age of the dogs she saw, giving your answer to 2 significant figures. (1 mark)

....

.... years

[Total: 3 marks]

Histograms

(1) The table gives information about the ages of children at a nursery.

Age (a years)	Frequency
$0 < a \leq 0.5$	8
$0.5 < a \leq 1$	12
$1 < a \leq 2$	16
$2 < a \leq 3$	30
$3 < a \leq 5$	28

Draw a histogram for the information shown in the table. (4 marks, ★★★★)

Work out the frequency density for each class.

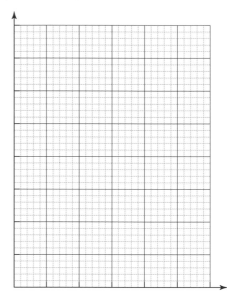

(2) The table and histogram show information about the wingspan of canaries in an aviary.

Use the histogram to complete the table. (3 marks, ★★★★)

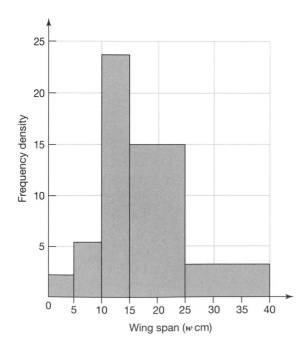

Wingspan (w cm)	Frequency
$0 < w \leq 5$	
$5 < w \leq 10$	6
$10 < w \leq 15$	24
$15 < w \leq 25$	30
$25 < w \leq 40$	

Cumulative frequency graphs

1 Tom receives a lot of junk mail through the post. He decides to keep a record of the amount of unwanted mail he receives over a period of two months. The table records the number of items per day and their corresponding frequency. (★★★)

Number of items of junk mail per day (n)	Frequency (days)	Cumulative frequency
$0 < n \leq 2$	21	
$2 < n \leq 4$	18	
$4 < n \leq 6$	10	
$6 < n \leq 8$	8	
$8 < n \leq 10$	3	
$10 < n \leq 12$	1	

a Complete the cumulative frequency column. (1 mark) ← This is the running total.

b On the grid, draw the cumulative frequency graph for the data in the table. (2 marks) ← Plot the upper value of the class interval against its cumulative frequency.

c Use your graph to find the median daily number of items of junk mail Tom received over the two months. (1 mark)

....

[Total: 4 marks]

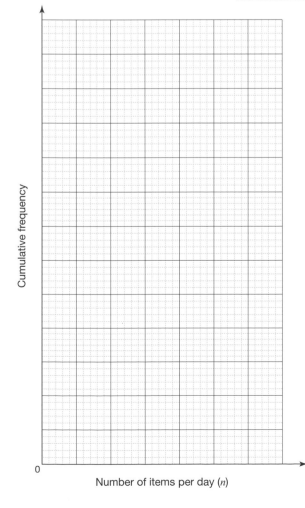

Cumulative frequency

Number of items per day (n)

(2) The cumulative frequency graph gives information about the masses of young emperor penguins in a zoo. (★★★)

a Find the interquartile range of the masses. (2 marks)

.... kg

You may use a ruler to draw lines on the graph to help you.

b Estimate the number of penguins with a mass above 10 kg. Show your workings. (2 marks)

....

[Total: 4 marks]

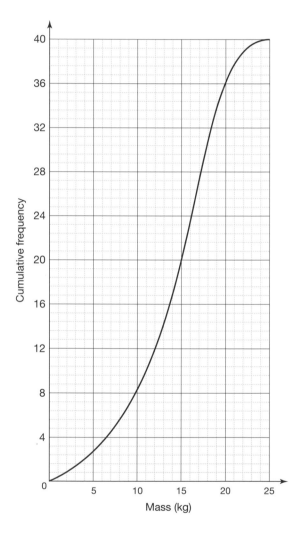

STRETCHIT!

The graph below has a point (25, 40).

Ruby says, 'What big penguins they've got at the zoo – 40 of them weigh 25 kg!'

Ruby is interpreting the graph incorrectly. Make sure you understand why.

Comparing sets of data

① Here are Chloe's scores in 15 spelling tests. (★★★)

10 12 17 17 15 18 17 16 14 15 14 12 17 19 18

> **NAILIT!**
>
> When comparing sets of data compare both an average (mean, median or mode) and a measure of spread (range or interquartile range).

a On the grid, draw a box plot for this information. (3 marks)

Sasha takes the same 15 spelling tests and her results are summarised below.

Her median score is 17.

The interquartile range for her scores is 2.

The range of her scores is 5.

b Compare Sasha and Chloe's scores. (2 marks)

....

....

....

....

....

[Total: 5 marks]

2 There were 100 male runners and 100 female runners taking part in a race. A cumulative frequency graph of the men's running times and a box plot of the women's running times are shown below. (★★★★★)

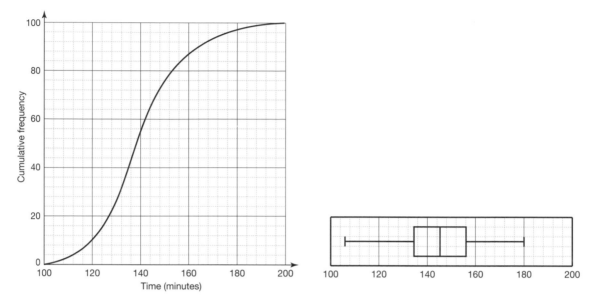

a Use the cumulative frequency graph graph to work out the following for the male runners

   i the median time (1 mark)

   .... minutes

   ii the interquartile range (1 mark)

   .... minutes

b Using both the graph and the box plot, compare the performances of the male and female runners. (2 marks)

   ....

   ....

   ....

[Total: 4 marks]

Scatter graphs

NAILIT!

Scatter graphs are used to investigate whether a correlation exists between two quantities. If the points show an upward slope to the right there is a positive correlation and if they show a downward slope to the right there is a negative correlation. The **line of best fit** demonstrates this. When drawing one, position it so that the points are evenly distributed on both sides.

1. A driving school advertises for students. The manager wants to find out whether the money spent on advertising affects the number of new students booking lessons. She collects the following data. (★★)

Advertising spend per week (£)	Number of new students per week from adverts
20	1
30	2
40	4
50	4
60	5
70	6

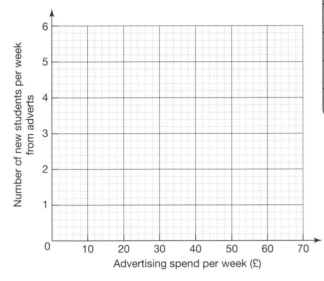

a Using the grid, draw a scatter graph to show this information. (2 marks)

b Draw a line of best fit on your graph. (1 mark)

c Describe the type of correlation. (1 mark)

....

[Total: 4 marks]

2. The scatter graph below shows the test marks in physics and maths for the same set of students. (★★)

a Draw the line of best fit. (1 mark)

b A student obtained a mark of 70% in the physics exam but was absent for the maths exam. Use your graph to predict the student's likely mark in the maths exam. (1 mark)

....

c Rhiannon scored 97% in physics. Explain why your line would not give an accurate estimate of her maths mark. (1 mark)

....

....

....

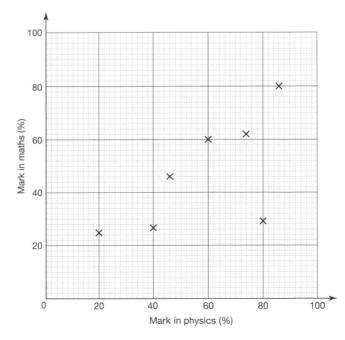

[Total: 3 marks]

Higher tier

Time: 1 hour 30 minutes

The maximum mark for this paper is 80.
The marks for questions are shown in brackets.

1   Which sequence is a geometric progression?

Circle your answer.

$1, \frac{1}{2}, \frac{1}{4}, \frac{1}{8}, \frac{1}{16}, ...$ 2, 3, 4, 5, 6, ...

2, 5, 8, 11, 14, ... $1, \frac{1}{3}, \frac{1}{4}, \frac{1}{5}, \frac{1}{6}, ...$

[1 mark]

2   Which statement about similar triangles is **not** correct?

Circle your answer.

All three angles of one triangle are equal to all three angles of the other triangle.

The ratio of pairs of corresponding sides is the same.

The triangles are identical.

One of the triangles is an enlargement of the other.

[1 mark]

3   Which expression is 260 written as the product of prime factors?

Circle your answer.

130 × 2   130 × 10 × 2

$2^2 \times 5 \times 13$ 13 × 5 × 4

[1 mark]

4   Circle the equation of a circle.

$y = 2x - 7$ $y = (x - 2)^2$

$y^2 = 25 - x^2$ $y = \frac{1}{x^2}$

[1 mark]

5   Work out $3\frac{1}{3} \times 1\frac{3}{8}$

Give your answer as a mixed number in its simplest form.

....

[2 marks]

6   Two spheres have radii in the ratio of 2:1

Circle the ratio of their volumes.

2:1   4:1

8:1   6:1

[1 mark]

7   n is an integer in the range $1 \leq n < 5$

The nth term of a sequence is $2n + 1$

The nth term of a different sequence is $n^2 + 1$

Find the value of the number that is in both sequences.

....

[3 marks]

8   Simplify $5^4 \times 5^{-1} \times 5$

Circle the answer.

5^4 5^5 5^3 125

[1 mark]

9   A sheet of paper has a length of 30 cm and a width of 20 cm.

Circle the area of the sheet in m².

0.6   0.06   6   60

[1 mark]

10

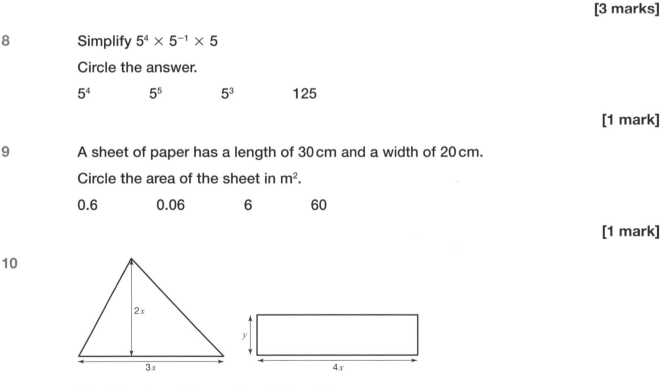

The triangle and the rectangle have the same area.

a   Show that $x(3x - 4y) = 0$

[3 marks]

b x and y are positive whole numbers.

Find two pairs of possible values for x and y.

....

[2 marks]

11   A bag contains red, green and blue counters.

	Red	Green	Blue
Number of counters	x	$x - 3$	$2x$

A counter is chosen at random.

The probability that the counter is green is $\dfrac{1}{5}$

Work out the probability that the counter is blue.

....

[3 marks]

12 $y \times 10^4 + y \times 10^2 = 36\,360$

Work out $y \times 10^4 - y \times 10^2$

Give your answer in standard form.

....

[2 marks]

13   The diagram shows the circle with equation $x^2 + y^2 = 25$

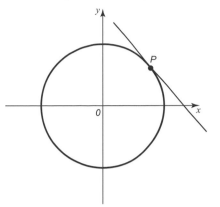

a Prove that the point with coordinates (4, 3) lies on the circle.

[2 marks]

b A tangent is drawn through point P(4, 3)

   Find the equation of this tangent.

....

[3 marks]

14   Rationalise the denominator and simplify $\dfrac{8}{3\sqrt{2}}$

....

[2 marks]

105

15   Simplify

a $(2x^2 y)^3$

....

[1 mark]

b $2x^{-3} \times 3x^4$

....

[1 mark]

c $\dfrac{15a^3 b}{3a^3 b^2}$

....

[1 mark]

16   x and y are positive integers where $y > x$.
        Tick whether each statement is true or false.
        Give a reason for your answer.

a $\dfrac{x}{y} < 1$

True ☐   False ☐

Reason ....

[1 mark]

b   The value of x^3 can never be -1

True ☐   False ☐

Reason ....

[1 mark]

c $x - y > 0$

True ☐   False ☐

Reason ....

[1 mark]

d $y^2 \geq x^2$

True ☐   False ☐

Reason ....

[1 mark]

17   Convert $0.1\dot{3}\dot{6}$ to a fraction in its lowest terms.

....

[3 marks]

18   Show that $\tan 30° + \sin 60°$ can be written as $\dfrac{5\sqrt{3}}{6}$

[3 marks]

19   Show that $(x + 1)(x + 1)(x + 1) = x^3 + 3x^2 + 3x + 1$

[2 marks]

20   Out of 300 students in a college who take one or more sciences from physics, chemistry and biology:

40% take all three sciences.

40 take only chemistry and biology. 20 take only physics and biology. 12 take only physics and chemistry.

The number who take only physics is equal to the number who take only chemistry.

20 take only biology.

a   Complete the Venn diagram.

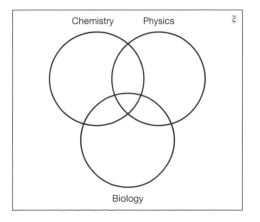

[2 marks]

b Work out the number of students who take only chemistry.

....

[2 marks]

c A student chosen at random takes biology.

Work out the probability that the student also takes chemistry.

....

[1 mark]

21

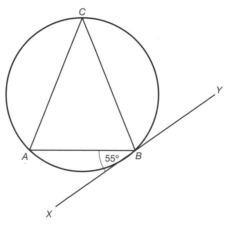

A, B, C are points on a circle.

ABC is an isosceles triangle with AC = BC

XY is a tangent drawn through point B.

Angle XBA = 55°

a Work out the size of angle ACB.

Give a reason for your answer.

....

Reason ....

[2 marks]

b Work out the size of angle *CBY*.

   Give reasons for each stage of your working.

....

[2 marks]

22   The graph shows information about how many pets students in a class have.

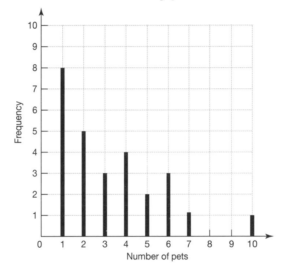

a Amy is a member of the class.

   She has the same number of pets as the mean number of pets for the class (to the nearest integer).

   How many pets does Amy have? Circle the correct answer.

   1   **3**   **4**   **5**

[2 marks]

b Jacob is in the same class.

   Jacob has the same number of pets as the median for the class.

   How many pets does Jacob have?

....

[2 marks]

23   *OAB* is a triangle.

M is the midpoint of *OB*.

N is the midpoint of *AB*.

P is the midpoint of *OA*.

R lies on line *AM* such that $AR = 2RM$.

$\overrightarrow{OA} = \mathbf{a}$ and $\overrightarrow{OB} = \mathbf{b}$

a Work out the following vectors in terms of **a** and **b**.

   i $\overrightarrow{AM}$

....

[1 mark]

   ii $\overrightarrow{AR}$

....

[1 mark]

b Show that R lies on *BP*.

[3 marks]

24   The attendance at a football match is 100 000, correct to 2 significant figures.

Work out the difference between the maximum and the minimum the attendance could be.

....

[2 marks]

25   Triangle *ABC* is drawn on a centimetre grid.

A is (2, 2)   *B* is (6, 2)   *C* is (1, 4)

a Triangle *ABC* is enlarged by scale factor $-\frac{1}{2}$, centre (0, 0).

On the grid, draw the triangle after the enlargement.

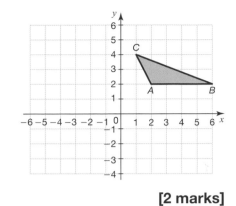

[2 marks]

b Triangle *ABC* is reflected in the line $y = x$.

State the number of invariant points on the perimeter of the triangle.

....

[1 mark]

26   Solve the simultaneous equations

$6x + 5y = 35$

$x - 2y = 3$

....

[4 marks]

27   Solve $x^2 \leq 2x + 15$

....

[3 marks]

28   Below is a sketch of the graph of $y = f(x)$

The graph has a turning point at (0, 6).

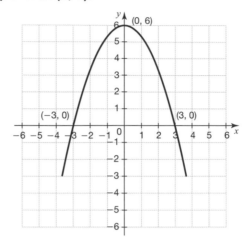

Write down the coordinates of the turning point of the curve with equation

a $y = f(x - 3)$

....

[1 mark]

b $y = -f(x)$

....

[1 mark]

29 OA is a straight line passing through the origin and point $A(3, 4)$.

A straight line BC passes through A and is perpendicular to line OA.

Find the equation of line BC.

Give your answer in the form $ax + by + c = 0$ where a, b and c are integers.

....

[4 marks]

Practice paper 2 (calculator)

Higher tier

Time: 1 hour 30 minutes

The maximum mark for this paper is 80.
The marks for questions are shown in brackets.

1   Circle the expression that is the same as $2x^2 + \dfrac{3x^2}{x} + x$

    $2x^2 + 4x$ $5x^2 + x$

    $3x^2 + 4x$ $9x$

                                                  [1 mark]

2   Circle the solutions to the equation $(x - 3)(x + 4) = 0$

    $x = -3$ or 4 $x = 3$ or 4

    $x = 3$ or -4 $x = 0$

                                                  [1 mark]

3   Circle the square root of $10\,000$.

    100 1000

    500 5000

                                                    [1 mark]

4   $\mathbf{a} = \begin{pmatrix} 2 \\ 1 \end{pmatrix}$ and $\mathbf{b} = \begin{pmatrix} -1 \\ 0 \end{pmatrix}$

Circle the vector that is equal to $3\mathbf{a} - 2\mathbf{b}$

    $\begin{pmatrix} 8 \\ 3 \end{pmatrix}$ $\begin{pmatrix} 4 \\ 3 \end{pmatrix}$

    $\begin{pmatrix} 5 \\ 3 \end{pmatrix}$ $\begin{pmatrix} 0 \\ 3 \end{pmatrix}$

                                                [1 mark]

5   A house increased in value from £150 000 in 2014 to £170 000 in 2016.

What percentage increase is this?

Give your answer to 1 decimal place.

....

                                                [2 marks]

6   Use your calculator to work out $19.85^2 - \sqrt{98.67} \div 4.67$

   a Write down your full calculator display.

....

                                                [1 mark]

   b Use approximations to check that the answer to part (a) is sensible.

      You **must** show your working.

....

                                                [1 mark]

7   The exterior angle of a regular polygon is 60°

Circle the name of the regular polygon.

Hexagon   Pentagon

Octagon   Triangle

[1 mark]

8   Molly picks a 4-digit number.

The first digit is not zero.

The second digit is 5.

The whole number is an even number.

How many different 4-digit numbers could she pick?

....

[3 marks]

9   y is inversely proportional to x.

Which graph correctly shows this?

Circle the correct letter.

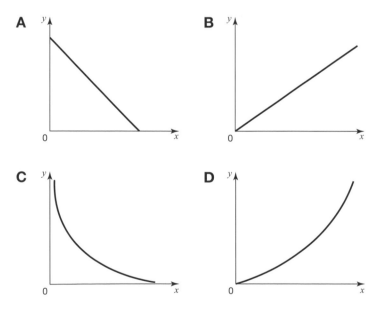

[1 mark]

10   Circle the statement about the straight-line graph that is correct.

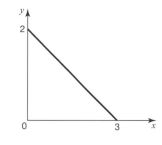

The gradient of the line is $-\dfrac{3}{2}$

The gradient of the line is $\dfrac{2}{3}$

The equation of the straight line is $y = -\dfrac{2}{3}x + 2$

The equation of the straight line is $y = \dfrac{2}{3}x - 2$

[1 mark]

11   Written as the product of its prime factors

$630 = 2 \times 3^2 \times 5 \times 7$

a Write 245 as the product of its prime factors.

....

[2 marks]

b Work out the value of the highest common factor of 630 and 245.

....

[1 mark]

12    $s = \dfrac{1}{2}at^2$

$a = 2.7 \times 10^8$

$t = 7.5 \times 10^{-3}$

Work out the value of s.

Give your answer to an appropriate degree of accuracy.

....

[3 marks]

13   Two circles have radii in the ratio 2:7

Circle the ratio of their areas.

2:7   4:14   4:49   1:3.5

[1 mark]

14   Here is some information about the number of pets each household in a certain street has.

One of the frequencies is missing.

Number of households	Frequency	Midpoint
0–2	22	1
3–5	12	4
6–8		7
9–11	3	10

The mean is estimated to be 3.25 using the midpoints.

Find the value of the missing frequency.

....

[3 marks]

15   a Factorise $10x^2 + 17x + 3$

....

[2 marks]

b Hence solve the equation $10x^2 + 17x + 3 = 0$

....

[2 marks]

16   The diagram shows a solid block of copper in the shape of a cuboid.

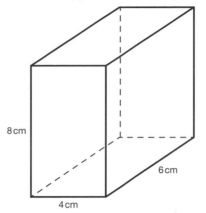

8 cm

6 cm

4 cm

a Find the volume of the block in cm³.

.... cm³

[1 mark]

b Find the volume of the block in m³.

   Give your answer in standard form.

.... m³

[1 mark]

c The density of copper is 8940 kg/m³.

   Find the mass of the block in kg.

   Give your answer correct to 2 decimal places.

.... kg

[1 mark]

d Use the formula

       force in N = mass in kg × 9.81

   to calculate the force produced by the block.

   Give your answer correct to 2 decimal places.

.... N

[1 mark]

e Use the formula

$$\text{pressure} = \frac{\text{force in N}}{\text{area in m}^2}$$

to find the maximum pressure the block could exert.

Give your answer in N/m² correct to 3 significant figures.

.... N/m²

[2 marks]

17   The waiting time in seconds for a call to be answered in a call centre is shown in the table.

Time for call to be answered (*t* seconds)	Frequency	Cumulative frequency
$0 \le t \le 10$	12	
$10 < t \le 20$	18	
$20 < t \le 30$	25	
$30 < t \le 40$	10	
$40 < t \le 50$	2	
$50 < t \le 60$	1	

a Complete the cumulative frequency column.

[2 marks]

b On the grid, draw the cumulative frequency graph.

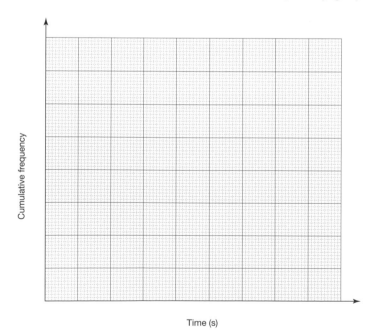

Time (s)

[2 marks]

c Find the median time for a call to be answered.

.... seconds

[1 mark]

18   *A* and *B* lie on the circumference of a circle, centre *O*.

The tangents at *A* and *B* meet at *P*.

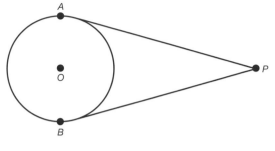

a   Prove that $AP = BP$

[2 marks]

b   Show that triangles *OAP* and *OBP* are congruent.

[1 mark]

19   One edge of a cube is increased in length by 3 cm.

A second edge is decreased in length by 2 cm.

The volume of the cuboid formed is 55 cm³ more than the volume of the cube.

Find the length of the side of the original cube.

.... cm

[5 marks]

20   Make y the subject of $ay + c = \dfrac{3 - y}{2}$

....

[3 marks]

21   An alloy consists of two metals, metal A and metal B in the ratio $3:5$.

Metal A costs £50 per kilogram and metal B costs £140 per kilogram.

Work out the cost of 500 g of the alloy.

....

[3 marks]

22   A ship sails due north from a port for 40 km. It then turns and travels on a bearing of 040° for 50 km.

a Calculate the direct distance of the ship from the port.

Give your answer correct to 1 decimal place.

.... km

[2 marks]

b Work out the bearing the ship must take for it to sail directly back to the port.

.... °

[3 marks]

23   During a spell of hot weather, the volume of water in a garden pond decreases due to evaporation.

The volume of water in the pond is initially $30\,m^3$.

The volume of water decreases at a rate of 2% per week.

a Find the volume of water in the pond after 4 weeks.

Give your answer correct to 3 significant figures.

.... m^3

[2 marks]

b If the rate of evaporation remains the same, after how many weeks will the volume of water first fall below half the initial volume?

.... weeks

[3 marks]

24   36 staff at a school are planning a night out.

$\frac{4}{9}$ of the staff are male.

Out of the male staff, 25% would like to go to the races and the rest would like to go for a meal.

Of the female staff, $\frac{1}{4}$ would like to go for a meal and the rest would like to go to a concert.

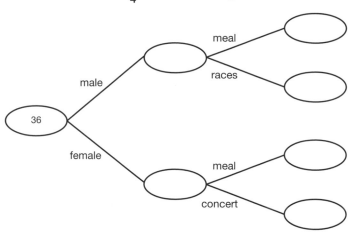

a Use this information to complete the frequency tree.

[2 marks]

b A member of staff is chosen at random. Work out the probability that this member of staff would prefer to go for a meal.

....

[2 marks]

25   Shade the region represented by these inequalities.

$x + 3y \leq 24$

$3x + y < 21$

$x \geq 2$

$y \leq 7$

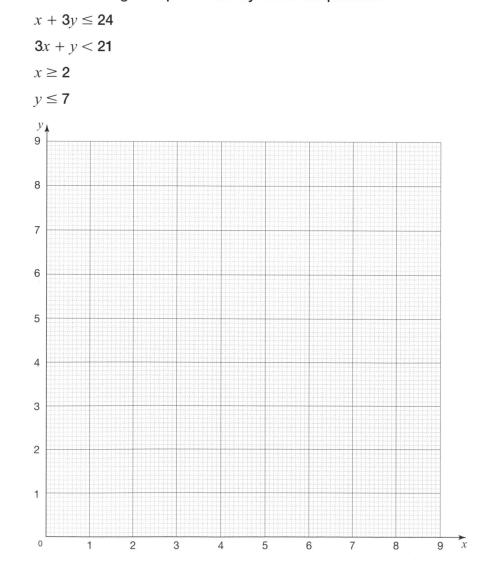

[5 marks]

26

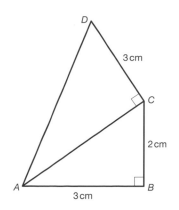

ABC and ACD are right-angled triangles.

Show that the length of AD is √22 cm.

[4 marks]

27   P, Q, R and S are points on a circle with centre O.

PR is a diameter of the circle.

Angle PSQ = 60°

Angle SQR = 21°

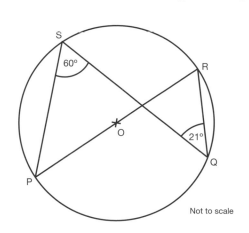

Not to scale

a Work out the size of angle PRQ giving reasons for your answer.

[2 marks]

b Work out the size of angle PRS giving reasons for your answer.

[2 marks]

Answers

Number

Integers, decimals and symbols

1 $\frac{1}{0.01}$ 0.1 $(0.1)^2$ $\frac{1}{1000}$ $(-1)^3$

2 **a** 35 **b** 0.01285 **c** −270 **d** 40

3 **a** 4644 **b** 4644 **c** 86 **d** 540

4 **a** 12.56 × 3.45 = 0.1256 × 345
   b $(-8)^2 > -64$ **c** 6 − 12 = 8 − 14
   d $(-7) \times (0) < (-7) \times (-3)$

Addition, subtraction, multiplication and division

1 **a** 76.765 **b** 201.646 **c** 91.33 **d** 10.564

2 **a** 1176 **c** 44.62 **e** 27
   b 2166 **d** 0.6572 **f** 63

3 **a** 1156 **b** 7.5 **c** 5.76

Using fractions

1 $\frac{2}{5} = \frac{16}{40} = \frac{30}{75} = \frac{50}{125}$

2 **a** $5\frac{1}{3}$ **b** $9\frac{7}{13}$

3 **a** $7\frac{1}{12}$ **b** $7\frac{1}{2}$ **c** $2\frac{9}{20}$

4 $\frac{5}{56}$ 5 $\frac{1}{2}$ $\frac{7}{12}$ $\frac{2}{3}$ $\frac{3}{4}$ $\frac{7}{8}$

Different types of number

1 **a** 7 **b** 49 **c** 2 **d** 6 **e** 6

2 **a** $3^2 \times 7 \times 11$ **b** 63 **c** 10395

3 441   4 5 minutes

Listing strategies

1 210 seconds   3 1100 students

2 5 friends   4 15 pairs

The order of operations in calculations

1 **a** Ravi has worked out the expression from left to right, instead of using BIDMAS. He should have performed the division and multiplication before the addition.
   b Correct answer: 40

2 **a** 122 **b** −3 **c** 40

3 **a** 6 **b** 14 **c** 8

Indices

1 **a** 10^6 **b** 10^8 **c** 10^6 **d** 10^3

2 **a** 1 **b** $\frac{1}{9}$ **c** 2 **d** 7

3 **a** $\frac{3}{2}$ **b** 16 **c** $\frac{1}{6}$ **d** 64

4 $x = 1.5$

Surds

1 **a** 5 **b** 30 **c** 18

2 $\frac{5\sqrt{3}}{4}$

3 $(2 + \sqrt{3})(2 - \sqrt{3}) = 4 - 2\sqrt{3} + 2\sqrt{3} - 3 = 1$

4 $a = 30$

5 $-\sqrt{5} - 7$

6 $\frac{1}{\sqrt{2}} + \frac{1}{4} = \frac{1 \times \sqrt{2}}{\sqrt{2} \times \sqrt{2}} + \frac{1}{4}$

   $= \frac{\sqrt{2}}{2} + \frac{1}{4}$

   $= \frac{2\sqrt{2}}{4} + \frac{1}{4}$

   $= \frac{1 + 2\sqrt{2}}{4}$

7 $\frac{2}{1 - \frac{1}{\sqrt{2}}} = \frac{2}{\frac{\sqrt{2}}{\sqrt{2}} - \frac{1}{\sqrt{2}}}$

   $= \frac{2}{\frac{\sqrt{2} - 1}{\sqrt{2}}}$

   $= \frac{2\sqrt{2}}{\sqrt{2} - 1}$

   $= \frac{2\sqrt{2}}{\sqrt{2} - 1} \times \frac{\sqrt{2} + 1}{\sqrt{2} + 1}$

   $= \frac{4 + 2\sqrt{2}}{2 - 1}$

   $= 4 + 2\sqrt{2}$

8 $\frac{3}{\sqrt{3}} + \sqrt{75} + (\sqrt{2} \times \sqrt{6}) = \frac{3\sqrt{3}}{3} + \sqrt{3 \times 25} + \sqrt{12}$

   $= \sqrt{3} + 5\sqrt{3} + \sqrt{3 \times 4}$

   $= \sqrt{3} + 5\sqrt{3} + 2\sqrt{3}$

   $= 8\sqrt{3}$

Standard form

1 **a** 2.55×10^{-3} **b** 1.006×10^{10} **c** 8.9×10^{-8}

2 **a** 6×10^{14} **c** 2×10^2 **e** 9×10^{-3}
   b 1.1×10^6 **d** 1×10^{-2}

3 2680   4 $a = 3.3$

Converting between fractions and decimals

1 **a** 0.55 **b** 0.375

2 **a** terminating **b** recurring **c** recurring

3   Let $x = 0.\dot{4}0\dot{2} = 0.402402402...$
      $1000x = 402.402402...$
      $1000x - x = 402.402402... - 0.402402402...$
      $999x = 402$
      $x = \frac{402}{999} = \frac{134}{333}$
      Hence $0.\dot{4}0\dot{2} = \frac{134}{333}$

4 $\frac{323}{495}$

Converting between fractions and percentages

1 **a** $\frac{7}{20}$ **b** $\frac{7}{100}$ **c** $\frac{19}{25}$ **d** $\frac{1}{8}$

2 **a** 20% **b** 68% **c** 250% **d** 17.5%

3 53.33% (to 2 d.p.)

4 $\frac{66}{90} = \frac{66}{90} = 73.3\%$ (to 1 d.p.)
   Jake did better in chemistry.

Fractions and percentages as operators

1 £34.79   4 **a** £14400 5 $\frac{14}{33}$

2 48   **b** £320

3 7040

Standard measurement units

1 175000 cm   2 17

3 1286 (to nearest whole number)

4 **a** 1.99×10^{-23} g (to 3 s.f.) **b** 1.99×10^{-26} kg (to 3 s.f.)

5 7.20×10^{-26} g (to 3 s.f.)

Rounding numbers

1 **a** 35 **c** 0 **e** 2
   b 101 **d** 0

2 **a** 34.88 **b** 34.877

3 **a** 12800 **b** 0.011 **c** 7×10^{-5}

4 **a** −0.00993 **b** 34.4 **c** 12300

Estimation

1 200   3 0.16   5 10.6

2 **a** 236.2298627 4 5   6 4
   b 240

7 **a** 5×10^{-28} kg
   b This will be an underestimate, as the mass of one electron has been rounded down.

Upper and lower bounds

1 $2.335 \leq l < 2.345$ kg

2 **a** **i** 2.472 **ii** 2.451 **b** 2.5 (to 2 s.f.)

3 34

Algebra

Simple algebraic techniques

1 **a** formula   **c** expression   **e** formula
   b identity   **d** identity
2 $x + 6x^2$
3 $y^3 - y = (1)^3 - 1 = 0$ so $y = 1$ is correct.
   $y^3 - y = (-1)^3 - (-1) = -1 + 1 = 0$ so $y = -1$ is correct.
4 **a** $10x$ **b** $4x^2 - 6x$ **c** $18x^2$
5 **a** 2   **b** $\frac{7}{8}$ **c** $-\frac{3}{2}$

Removing brackets

1 **a** $24x - 56$ **b** $-6x + 12$
2 **a** $3x + 9$ **c** $10a^2b - 5ab^2$
   b $8xy + 6x - 2y$ **d** $2x^3y^3 + 3x^2y^4$
3 **a** $m^2 + 5m - 24$ **c** $9x^2 - 6x + 1$
   b $8x^2 + 26x - 7$ **d** $6x^2 + xy - y^2$
4 **a** $x^2 + 7x + 10$ **c** $x^2 - 6x - 7$
   b $x^2 - 16$ **d** $15x^2 + 14x + 3$
5 **a** $x^3 + 6x^2 + 5x - 12$ **b** $18x^3 - 63x^2 + 37x + 20$

Factorising

1 **a** $5x(5x - y)$ **b** $2\pi(2r^2 + 3x)$ **c** $6ab^2(a^2 + 2)$
2 **a** $(3x + 1)(3x - 1)$ **b** $4(2x + 1)(2x - 1)$
3 **a** $(a + 4)(a + 8)$ **b** $(p - 6)(p - 4)$
4 **a** $a(a + 12)$ **b** $(b + 3)(b - 3)$ **c** $(x - 5)(x - 6)$
5 **a** $(3x + 8)(x + 4)$ **b** $(3x + 13)(x - 1)$ **c** $(2x - 5)(x + 2)$
6 $\frac{2}{(x - 3)}$ 7 $\frac{2x - 1}{4x + 1}$

Changing the subject of a formula

1 $T = \frac{PV}{nR}$ 3 $a = \frac{v - u}{t}$ 5 $v = \sqrt{\frac{2E}{m}}$
2 $y = \frac{1 - 4x}{2}$ 4 $x - 5(y + m)$
6 **a** $r = \sqrt{\frac{3V}{\pi h}}$ **b** 3.45 cm (to 2 d.p.)
7 **a** $x = \frac{y + 9}{3}$ **b** 4
8 $x = \frac{3y - 2}{a + 1}$
9 **a** $c = \frac{b}{a}$ **b** upper bound for $c = 1.18$ (to 3 s.f.)
                lower bound for $c = 1.11$ (to 3 s.f.)

Solving linear equations

1 **a** $x = 7$ **d** $x = 32$ **g** $x = -2$
   b $x = 5$ **e** $x = 25$ **h** $x = 84$
   c $x = 4$ **f** $x = -9$
2 $x = \frac{2}{3}$
3 **a** $x = \frac{1}{2}$ **b** $x = -\frac{8}{5}$

Solving quadratic equations using factorisation

1 **a** $(x - 3)(x - 4)$ **b** $x = 3$ or $x = 4$
2 **a** $(2x - 1)(x + 3)$ **b** $x = \frac{1}{2}$ or $x = -3$
3 $x = -2$ or $x = 6$
4 **a** $x(x - 8) - 7 = x(5 - x)$ **b** $x = -\frac{1}{2}$ or $x = 7$
   $x^2 - 8x - 7 = 5x - x^2$
   $2x^2 - 13x - 7 = 0$
5 $x = 2$ cm

Solving quadratic equations using the formula

1 **a** $\frac{3}{x + 7} = \frac{2 - x}{x + 1}$ **b** $x = 1.20$ or -9.20 (to 2 d.p.)
   $3(x + 1) = (2 - x)(x + 7)$
   $3x + 3 = 2x + 14 - x^2 - 7x$
   $3x + 3 = -x^2 - 5x + 14$
   $x^2 + 8x - 11 = 0$
2 $x = 2.78$ cm (to 2 d.p.)   3 $x = 3.30$ or -0.30 (to 2 d.p.)

Solving simultaneous equations

1 $x = 2$ and $y = 3$

2 **a** 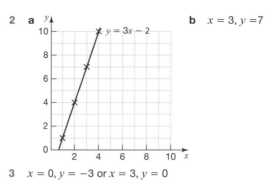   **b** $x = 3, y = 7$

3 $x = 0, y = -3$ or $x = 3, y = 0$

Solving inequalities

1 **a** $x \geq -9$ **b** $x < -12$
2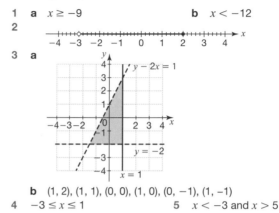
3 **a**

   b $(1, 2), (1, 1), (0, 0), (1, 0), (0, -1), (1, -1)$
4 $-3 \leq x \leq 1$ 5 $x < -3$ and $x > 5$

Problem solving using algebra

1 42 m²   2 cost of adult ticket = £7.50
            cost of child ticket = £4
3 **a** 16 years   **b** 9 years

Use of functions

1 **a** 19   **b** $x = -1$
2 **a** $(x - 6)^2$ **b** $x^2 - 6$
3 **a** ± 3 **b** $2x + 5$
4 $f^{-1}(x) = \sqrt{\frac{x - 3}{5}}$

Iterative methods

1 Let $f(x) = 2x^3 - 2x + 1$
   $f(-1) = 2(-1)^3 - 2(-1) + 1 = 1$
   $f(-1.5) = 2(-1.5)^3 - 2(-1.5) + 1 = -2.75$
   There is a sign change of $f(x)$, so there is a solution between
   $x = -1$ and $x = -1.5$.
2 $x_1 = 0.1121111111$
   $x_2 = 0.1125202246$
   $x_3 = 0.1125357073$
3 **a** $x_4 = 1.5213705 \approx 1.521$ (to 3 d.p.)
   b Checking value of $x^3 - x - 2$ for $x = 1.5205, 1.5215$:
       When $x = 1.5205$ $f(1.5205) = -0.0052$
           $x = 1.5215$ $f(1.5215) = 0.0007$
   Since there is a change of sign, the root is 1.521 correct to
   3 decimal places.
4 **a i**

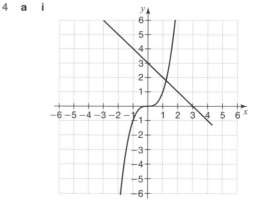

Answers

ii There is a root of $x^3 + x - 3 = 0$ where the graphs of $y = x^3$ and $y = 3 - x$ intersect. The graphs intersect once so there is one real root of the equation $x^3 + x - 3 = 0$.

b $x_1 = 1.216440399$
$x_2 = 1.212725591$
$x_3 = 1.213566964$
$x_4 = 1.213376503$
$x_5 = 1.213419623$
$x_6 = 1.213409861 = 1.2134$ (to 4 d.p.)

Equation of a straight line

1 A
2 **a** $-\frac{4}{3}$ **b** $y = -\frac{1}{2}x + \frac{7}{2}$ **c** $y = 2x + 1$
3 (3.8, 11.4) (to 1 d.p.)

Quadratic graphs

1 **a** $x = -0.3$ or -3.7 (to 1 d.p.)
  b

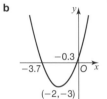

2 $a = 5$, $b = -2$ and $c = -10$
3 $a = 2$, $b = 3$ and $c = -15$

Recognising and sketching graphs of functions

1

Equation	Graph
$y = x^2$	B
$y = 2^x$	D
$y = \sin x°$	E
$y = x^3$	C
$y = x^2 - 6x + 8$	A
$y = \cos x°$	F

2 **a**

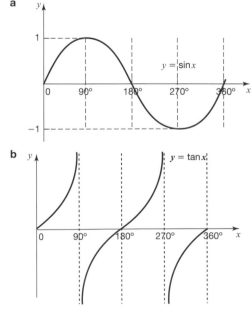

b

3 $\theta = 70.5°$ or $289.5°$ (to 1 d.p.)

Translations and reflections of functions

1 **a**

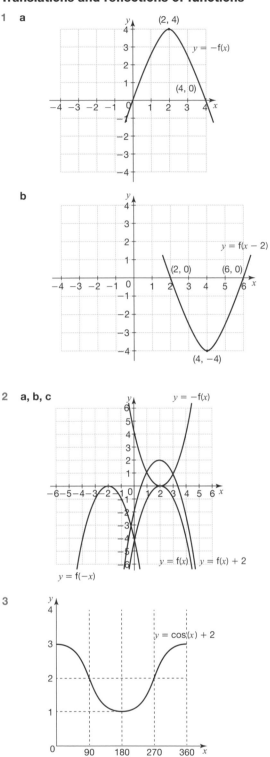

b

2 **a, b, c**

3

Equation of a circle and tangent to a circle

1 **a** 5
  b 7
  c 2
2 radius of the circle = $\sqrt{21} = 4.58$
  distance of the point (4, 3) from the centre of the circle (0, 0) = $\sqrt{16 + 9} = \sqrt{25} = 5$
  This distance is greater than the radius of the circle, so the point lies outside the circle.
3 **a** $\sqrt{74}$
  b $x^2 + y^2 = 74$ **c** $y = -\frac{5}{7}x + \frac{74}{7}$

Real-life graphs

1 **a** $1\,\text{m/s}^2$
   b 225 m
2 **a** The graph is a straight line starting at the origin, so this represents constant acceleration from rest of $\frac{15}{6} = 2.5\,\text{m/s}^2$.
   b The gradient decreases to zero, so the acceleration decreases to zero.
   c 118 m (to nearest integer); 117 m is also acceptable
   d It will be a slight underestimate, as the curve is always above the straight lines forming the tops of the trapeziums.

Generating sequences

1 **a** **i** $\frac{1}{2}$ **ii** 243 **iii** 21
   b 14, 1
2 −3, −11
3 **a** 25, 36
   b 15, 21
   c 8, 13

The *n*th term

1 **a** nth term $= 4n − 2$
   b nth term $= 4n − 2 = 2(2n − 1)$
      2 is a factor, so the nth term is divisible by 2 and therefore is even.
   c 236 is not a term in the sequence.
2 **a** 5
   b −391
   c n^2 is always positive, so the largest value $9 − n^2$ can take is 8 when $n = 1$. All values of n above 1 will make $9 − n^2$ smaller than 8. So 10 cannot be a term.
3 nth term $− n^2 − 3n + 3$

Arguments and proofs

1 The only prime number that is not odd is 2, which is the only even prime number.
   Hence, statement is false because 2 is a prime number that is not odd.
2 **a** true: $n = 1$ is the smallest positive integer and this would give the smallest value of $2n + 1$ which is 3.
   b true: 3 is a factor of $3(n + 1)$ so $3(n + 1)$ must be a multiple of 3.
   c false: $2n$ is always even and subtracting 3 will give an odd number.
3 Let first number $= x$ so next number $= x + 1$
   Sum of consecutive integers $= x + x + 1 = 2x + 1$
   Regardless of whether x is odd or even, $2x$ will always be even as it is divisible by 2.
   Hence $2x + 1$ will always be odd.
4 $(2x − 1)^2 − (x − 2)^2$
   $= 4x^2 − 4x + 1 − (x^2 − 4x + 4)$
   $= 4x^2 − 4x + 1 − x^2 + 4x − 4$
   $= 3x^2 − 3$
   $= 3(x^2 − 1)$
   The 3 outside the brackets shows that the result is a multiple of 3 for all integer values of x.
5 Let two consecutive odd numbers be $2n − 1$ and $2n + 1$.
   $(2n + 1)^2 − (2n − 1)^2$
   $= (4n^2 + 4n + 1) − (4n^2 − 4n + 1)$
   $= 8n$
   Since 8 is a factor of $8n$, the difference between the squares of two consecutive odd numbers is always a multiple of 8.
   (If you used $2n + 1$ and $2n + 3$ for the two consecutive odd numbers, difference of squares $= 8n + 8 = 8(n + 1)$.)

Ratio

Introduction to ratios

1 30
2 £7500, £8500, £9000
3 210 acres
4 $x = \frac{5}{7}$
5 144

Scale diagrams and maps

1 5 km
2 **a** 0.92 km **b** 0.12 km
3 1 : 800 4 1 : 200000

Percentage problems

1 10%
2 83.3%
3 £18000
4 £16250
5 £896

Direct and inverse proportion

1 **a** $P = kT$ **b** 74074 pascals (to nearest whole number)
2 £853 (to nearest whole number)
3 **a** $c = \frac{36}{h}$ **b** 2.4
4 **a** €402.50 **b** £72.07 (to nearest penny) **c** £2.50

Graphs of direct and inverse proportion and rates of change

1 B 2 B
3 **a**
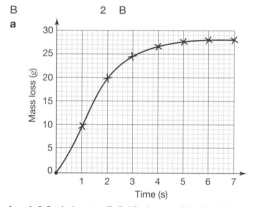
   b **i** 9.8 g/minute **ii** 0.16 g/second (to 2 d.p.)

Growth and decay

1 **a** 178652 **b** 5 years
2 £12594
3 0.1 (to 1 s.f.)

Ratios of lengths, areas and volumes

1 **a** 3.375 or $\frac{27}{8}$ **c** 133 cm³ (to nearest whole number)
   b 22.5 cm²
2 $h = 15$ cm (to nearest cm)
3 **a** **i** 9 cm **ii** 4.5 cm **b** 4 : 1

Gradient of a curve and rate of change

1 **a** $\frac{2}{3}\,\text{m/s}^2$ **c** $0.37\,\text{m/s}^2$
   b $0.26\,\text{m/s}^2$ **d** 34 s

Converting units of areas and volumes, and compound units

1 $500\,\text{N/m}^2$
2 $25\,000\,\text{N/m}^2$
3 107 g (to nearest g)
4 He has worked out the area in m² by dividing the area in cm² by 100, which is incorrect.
   There are $100 × 100 = 10000$ cm² in 1 m², so the area should have been divided by 10000.
   Correct answer:
   area in m² $= \frac{9018}{10000}$
   $= 0.9018$
   $= 0.90\,\text{m}^2$ (to 2 d.p.)
5 72 km/h

Answers

Geometry and measures

2D shapes

1 **a** true   **c** true   **e** true
   b false   **d** true   **f** false (this would be true only for a regular pentagon)

2 **a** rhombus   **c** equilateral triangle
   b parallelogram   **d** kite

Constructions and loci

1

2
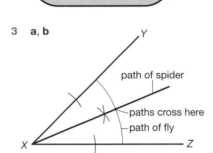

3 **a, b**
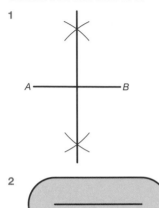

path of spider
paths cross here
path of fly

Properties of angles

1 **a** angle ACB = angle BAC = 30° (base angles of isosceles triangle ABC, since AB and BC are equal sides of a rhombus)
   b angle AOB = 90° (diagonals of a rhombus intersect at right angles)
   c angle ABO = 180 − (90 + 30) = 60° (angle sum of a triangle)
      angle BDC = angle ABO = 60° (alternate angles between parallel lines AB and DC)

2 angle $BAC = \frac{(180 - 36)}{2} = 72°$ (angle sum of a triangle and base angles of an isosceles triangle)
   angle BDC = 180 − 90 = 90° (angle sum on a straight line)
   angle ABD = 180 − (90 + 72) = 18° (angle sum of a triangle)

3 **a** $x = 30°$
   b If lines AB and CD are parallel, the angles $4x$ and $3x + 30$ would be corresponding angles, and so equal.
      $4x = 4 \times 30 = 120°$
      $3x + 30 = 3 \times 30 + 30 = 120°$
      These two angles are equal so lines AB and CD are parallel.

4 $x = 90 + 72 = 162°$

Congruent triangles

1 BD common to triangles ABD and CDB
   angle ADB = angle CBD (alternate angles)
   angle ABD = angle CDB (alternate angles)
   Therefore triangles ABD and CDB are congruent (ASA).
   Hence angle BAD = angle BCD

2 Draw the triangle and the perpendicular from A to BC.
   $AX = AX$ (common)
   $AB = AC$ (triangle ABC is isosceles)
   angle $AXB = AXC = 90°$ (given)
   Therefore triangles ABX and ACX are congruent (RHS).
   Hence $BX = XC$, so X bisects BC.

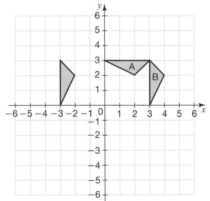

3 $OQC = 90°$ (corresponding angles), so $PB = OQ$ (perpendicular distance between 2 parallel lines)
   $AP = PB$ (given), so $AP = OQ$
   $PO = QC$ (Q is the midpoint of BC)
   angle ABC = angle APO = angle OQC
   = 90° (OQ is parallel to AB and OP parallel to BC)
   Therefore triangles AOP and OCQ are congruent (SAS).

Transformations

1 translation of $\begin{pmatrix} -7 \\ -5 \end{pmatrix}$

2
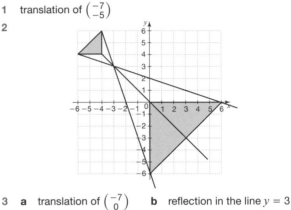

3 **a** translation of $\begin{pmatrix} -7 \\ 0 \end{pmatrix}$ **b** reflection in the line $y = 3$
   c rotation of 90° clockwise about (0, 1)

Invariance and combined transformations

1 **a** 1
   b **i** invariant point (3, 3)

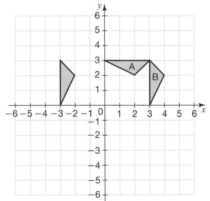

   ii rotation 90° anticlockwise about the point (3, 3)

2 **a** The shaded triangle is the image after the two transformations.

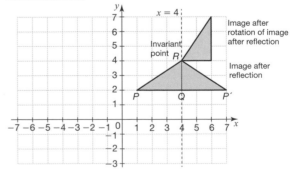

   b invariant point is R (4, 4)

3D shapes

1 **a** G   **c** A, H   **e** C
   b B   **d** B   **f** A, H

Parts of a circle

1 a radius   c chord
   b diameter   d arc
2 a minor sector   c major sector
   b major segment   d minor segment

Circle theorems

1 angle $OTB = 90°$ (angle between tangent and radius)
angle $BOT = 180 - (90 + 28) = 62°$
(angle sum in a triangle)
angle $AOT = 180 - 62 = 118°$
(angle sum on a straight line)
$AO = OT$ (radii), so triangle AOT is isosceles
angle $OAT = \frac{(180 - 118)}{2} = 31°$ (angle sum in a triangle)

2 a angle $ACB = 30°$ (angle at centre twice angle at circumference)

   b angle BAC = angle $CBX = 70°$ (alternate segment theorem)

   c $OA = OB$ (radii), so triangle AOB is isosceles
angle $AOB = 60°$, so triangle AOB is equilateral
angle $OAB = 60°$ (angle of equilateral triangle)
angle $CAO = 70 - 60 = 10°$

Projections

1

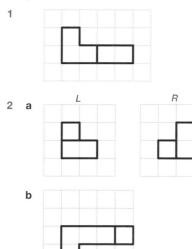

2 a

b

3

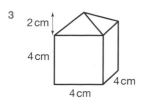

Bearings

1 230°
2

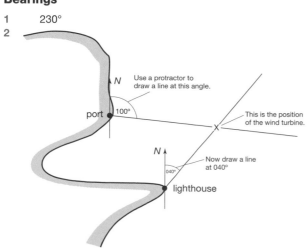

Pythagoras' theorem

1 76 m (to nearest m)
2 9.8 cm (to 1 d.p.)
3 a 9.1 cm (to 2 d.p.)
   b 48.76 cm² (to 2 d.p.)

Area of 2D shapes

1 a   i 5.70 cm (to 2 d.p.)
      ii 19.80 cm² (to 2 d.p.)
   b 50.65 cm²
2 a 9 cm²   b 6 cm²
3 a 27π cm²   b 18π + 6 cm

Volume and surface area of 3D shapes

1 a 3.975 m²   b 6.36 m³ (to 2 d.p.)
2 5 glasses
3 a 7.4 cm (to 1 d.p.)
   b 3.8 cm (to 1 d.p.)
4 0.64 cm

Trigonometric ratios

1 a 6.0 cm (to 1 d.p.)
   b 36.9° (to 1 d.p.)
2 44.4° (to 1 d.p.)
3 9.4 cm (to 1 d.p.)
4 21.8° (to 1 d.p.)

Exact values of sin, cos and tan

1

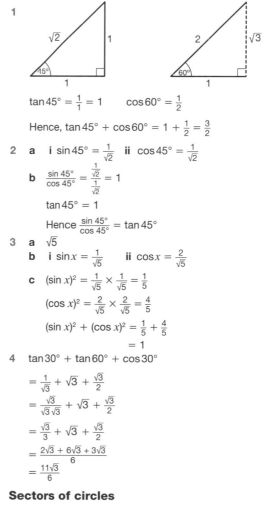

$\tan 45° = \frac{1}{1} = 1$ $\cos 60° = \frac{1}{2}$

Hence, $\tan 45° + \cos 60° = 1 + \frac{1}{2} = \frac{3}{2}$

2 a   i $\sin 45° = \frac{1}{\sqrt{2}}$ ii $\cos 45° = \frac{1}{\sqrt{2}}$

   b $\frac{\sin 45°}{\cos 45°} = \frac{\frac{1}{\sqrt{2}}}{\frac{1}{\sqrt{2}}} = 1$

     $\tan 45° = 1$

     Hence $\frac{\sin 45°}{\cos 45°} = \tan 45°$

3 a $\sqrt{5}$
   b   i $\sin x = \frac{1}{\sqrt{5}}$ ii $\cos x = \frac{2}{\sqrt{5}}$

   c $(\sin x)^2 = \frac{1}{\sqrt{5}} \times \frac{1}{\sqrt{5}} = \frac{1}{5}$

     $(\cos x)^2 = \frac{2}{\sqrt{5}} \times \frac{2}{\sqrt{5}} = \frac{4}{5}$

     $(\sin x)^2 + (\cos x)^2 = \frac{1}{5} + \frac{4}{5}$
                       $= 1$

4 $\tan 30° + \tan 60° + \cos 30°$

   $= \frac{1}{\sqrt{3}} + \sqrt{3} + \frac{\sqrt{3}}{2}$

   $= \frac{\sqrt{3}}{\sqrt{3}\sqrt{3}} + \sqrt{3} + \frac{\sqrt{3}}{2}$

   $= \frac{\sqrt{3}}{3} + \sqrt{3} + \frac{\sqrt{3}}{2}$

   $= \frac{2\sqrt{3} + 6\sqrt{3} + 3\sqrt{3}}{6}$

   $= \frac{11\sqrt{3}}{6}$

Sectors of circles

1 34° (to nearest degree)
2 364.4 cm²
3 a 1.67 cm (to 3 s.f.)

   b 1 : 1.04

4 a length of arc $AB = \frac{\theta}{360} \times 2\pi r$

$$5.4 = \frac{\theta}{360} \times 2\pi \times 6$$

$$\theta = 51.5662$$

Area of sector $AOB = \frac{\theta}{360} \times 2\pi r^2$

$$= \frac{51.5662}{360} \times \pi \times 6^2$$

$$= 16.2 \text{ cm}^2$$

Note that both a and b are equal to the radius r of the circle.

b area of triangle AOB

$$= \tfrac{1}{2} a\, b \sin c$$

$$= \tfrac{1}{2} \times 6 \times 6 \sin 51.5662$$

$$= 14.09988 \text{ cm}^2$$

area of shaded segment

$= $ area of sector $-$ area of triangle

$= 16.2 - 14.1$

$= 2.1\text{cm}^2$ (correct to 1 decimal place)

Sine and cosine rules

1 a $225\,\text{cm}^2$

b $\frac{4}{5}$

c $18.0\,\text{cm}$ (to 3 s.f.)

2 a $18.6\,\text{cm}$ (to 3 s.f.)

b $92.4\,\text{cm}^2$ (to 3 s.f.)

3 $\frac{2 + 6\sqrt{2}}{17}$

Vectors

1 a $\binom{2}{1}$ **b** $\binom{-22}{9}$

2 a $\mathbf{b} - \mathbf{a}$ **b** $\frac{3}{5}(\mathbf{b} - \mathbf{a})$

c $\overrightarrow{OQ} = \frac{2}{5}\overrightarrow{OA} = \frac{2}{5}\mathbf{a}$

$\overrightarrow{QP} = \overrightarrow{QA} + \overrightarrow{AP}$

$= \frac{3}{5}\mathbf{a} + \frac{3}{5}(\mathbf{b} - \mathbf{a})$

$= \frac{3}{5}\mathbf{a} + \frac{3}{5}\mathbf{b} - \frac{3}{5}\mathbf{a}$

$= \frac{3}{5}\mathbf{b}$

As $\overrightarrow{QP} = \frac{3}{5}\mathbf{b}$ and $\overrightarrow{OB} = \mathbf{b}$ they both have the same vector part and so are parallel.

3 a $\overrightarrow{BC} = \overrightarrow{BA} + \overrightarrow{AC}$

$= -3\mathbf{b} + \mathbf{a}$

$= \mathbf{a} - 3\mathbf{b}$

b $\overrightarrow{PB} = \frac{1}{3}\overrightarrow{AB} = \mathbf{b}$

$\overrightarrow{PM} = \overrightarrow{PB} + \overrightarrow{BM}$

$= \overrightarrow{PB} + \frac{1}{2}\overrightarrow{BC}$

$= \mathbf{b} + \frac{1}{2}(\mathbf{a} - 3\mathbf{b})$

$= \frac{1}{2}\mathbf{a} - \frac{1}{2}\mathbf{b}$

$= \frac{1}{2}(\mathbf{a} - \mathbf{b})$

$\overrightarrow{MD} = \overrightarrow{MC} + \overrightarrow{CD}$

$= \frac{1}{2}\overrightarrow{BC} + \overrightarrow{CD}$

$= \frac{1}{2}(\mathbf{a} - 3\mathbf{b}) + \mathbf{a}$

$= \frac{3}{2}\mathbf{a} - \frac{3}{2}\mathbf{b}$

$= \frac{3}{2}(\mathbf{a} - \mathbf{b})$

Both $\overrightarrow{PM}$ and $\overrightarrow{MD}$ have the same vector part $(\mathbf{a} - \mathbf{b})$ so they are parallel. Since they both pass through M, they are parts of the same line, so PMD is a straight line.

Probability

The basics of probability

1 a

Dice 1

Dice 2		1	2	3	4	5	6
	1	2	3	4	5	6	7
	2	3	4	5	6	7	8
	3	4	5	6	7	8	9
	4	5	6	7	8	9	10
	5	6	7	8	9	10	11
	6	7	8	9	10	11	12

b $\frac{1}{36}$ **c** $\frac{5}{12}$ **d** 7

2 a 21 chocolates **b** $\frac{3}{7}$

3

Bethany

Amy		1	2	3	4	5	6
	1	1, 1	1, 2	1, 3	1, 4	1, 5	1, 6
	2	2, 1	2, 2	2, 3	2, 4	2, 5	2, 6
	3	3, 1	3, 2	3, 3	3, 4	3, 5	3, 6
	4	4, 1	4, 2	4, 3	4, 4	4, 5	4, 6
	5	5, 1	5, 2	5, 3	5, 4	5, 5	5, 6
	6	6, 1	6, 2	6, 3	6, 4	6, 5	6, 6

a $\frac{1}{6}$ **b** $\frac{5}{12}$

Probability experiments

1 a He is wrong because 100 spins is a very small number of trials. To approach the theoretical probability you would have to spin many more times. Only when the number of spins is extremely large will the frequencies start to become similar.

b $\frac{11}{50}$ **c** 95

2 a $x = 0.08$ **b** 0.24 **c** 16

The AND and OR rules

1 a $\frac{9}{169}$ **b** $\frac{3}{169}$ **c** $\frac{1}{52} \times \frac{1}{52} = \frac{1}{2704}$

2 a When an event has no effect on another event, they are said to be independent events. Here the colour of the first marble has no effect on the colour of the second marble.

b $\frac{9}{100}$ **c** $\frac{3}{10}$

3 a $\frac{9}{140}$ **b** $\frac{6}{35}$

Tree diagrams

1 a $\frac{2}{15}$ **b** $\frac{8}{15}$

2 a

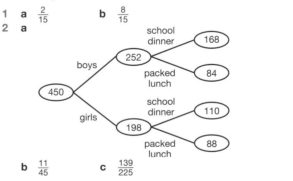

b $\frac{11}{45}$ **c** $\frac{139}{225}$

3 a 0.179 (to 3 d.p.) **b** 0.238 (to 3 d.p.) **c** 0.131 (to 3 d.p.)

Venn diagrams and probability

1 a 9, 8 **c** 1, 3, 4, 10, 12, 15

b 1, 2, 3, 5, 7, 8, 9, 12, 15 **d** 4, 10

2 a

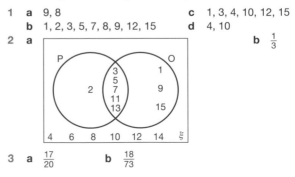

b $\frac{1}{3}$

3 a $\frac{17}{20}$ **b** $\frac{18}{73}$

Statistics

Sampling

1 a Ling's, as he has a larger sample so it is more likely to represent the whole population (i.e. students at the school).

  b 22

2 The sample should be taken randomly, with each member of the population having an equal chance of being chosen.

  The sample size should be large enough to represent the population, since the larger the sample size, the more accurate the results.

3 a 173

  b 25

Two-way tables and pie charts

1 a

	Coronation Street	EastEnders	Emmerdale	Total
Boys	12	31	20	63
Girls	18	12	7	37
Total	30	43	27	100

  b $\frac{27}{100}$

  c $\frac{12}{37}$

2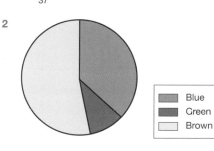

- Blue
- Green
- Brown

3 a 11 students
  b 23 students

Line graphs for time series data

1 a An upward trend, rising slowly at first, then rising quickly, then continuing to rise more slowly.

  b Mode, because it shows that June is the month when the environment offers the largest sample of insects to study. The median would give May, when there are also a lot of insects.

2 a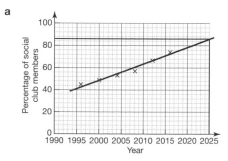

  b An upward trend – people are more likely to assume it should be an option.

  c 86% (or a close value)

  d Future data may change so that the line of best fit is no longer accurate. Also, the relationship may not be best represented by a straight line, but by a curve.

Averages and spread

1 a 32

  b 2.7 (to 1 d.p.)

  c 3

2 66.5 (to 1 d.p.)

3 a

Age (t years)	Frequency	Mid-interval value	Frequency × mid-interval value
$0 < t \le 4$	8	2	16
$4 < t \le 8$	10	6	60
$8 < t \le 12$	16	10	160
$12 < t \le 14$	1	13	13

  b 35

  c 7.1 years (to 2 s.f.)

Histograms

1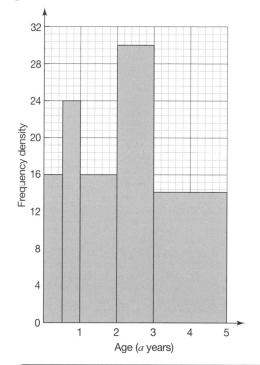

2

Wingspan (w cm)	Frequency
$0 < w \le 5$	2
$5 < w \le 10$	6
$10 < w \le 15$	24
$15 < w \le 25$	30
$25 < w \le 40$	9

Cumulative frequency graphs

1 a

Number of items of junk mail per day (n)	Frequency	Cumulative frequency
$0 < n \le 2$	21	21
$2 < n \le 4$	18	39
$4 < n \le 6$	10	49
$6 < n \le 8$	8	57
$8 < n \le 10$	3	60
$10 < n \le 12$	1	61

b

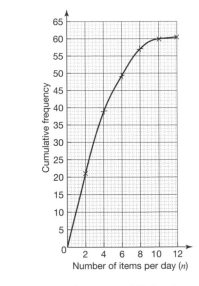

c median = value at frequency of 30.5 = 3

2 a 6.5 kg

b 32 penguins

Comparing sets of data

1 a

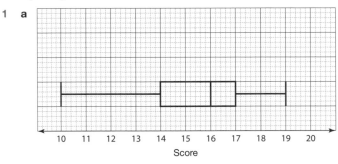

b Sasha's median score is higher (17 compared to 16). The IQR for Sasha is 2 compared to Chloe's 3. The range for Sasha is 5 compared to Chloe's 9. Both these are measures of spread, which means that Sasha's scores are less spread out (i.e. more consistent).

2 a   i 138 minutes   **ii** 22 minutes

b On average the men were faster as the median is higher for the men.

The variation in times was greater for the men as their range was greater, although the spread of the middle half of the data (the interquartile range) was slightly greater for the women.

Scatter graphs

1 a, b

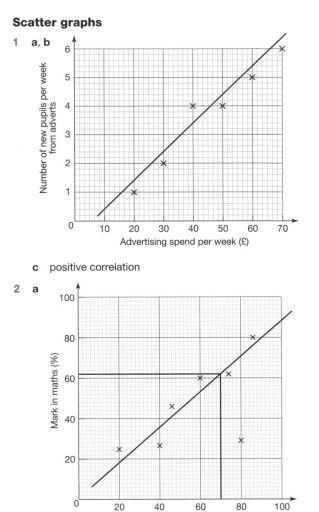

c positive correlation

2 a

b 62%

c The data only goes up to a score of 86% in physics, and the score at 80% in physics is an outlier, so the line may not be accurate for higher scores in physics.

For answers to the practice papers, visit:
www.scholastic.co.uk/gcse